Coastal Groundwater System Changes in Response to Large-Scale Land Reclamation

COASTAL GROUNDWATER SYSTEM CHANGES IN RESPONSE TO LARGE-SCALE LAND RECLAMATION

HAIPENG GUO
AND
JIU J. JIAO

Nova Science Publishers, Inc.
New York

For permission to use material from this book please contact us:
Telephone 631-231-7269; Fax 631-231-8175
Web Site: http://www.novapublishers.com

LIBRARY OF CONGRESS CATALOGING-IN-PUBLICATION DATA

ISBN: 978-1-60692-218-7

Available upon request

Published by Nova Science Publishers, Inc. New York

CONTENTS

PREFACE

Most large urban centers lie in coastal regions, which are home to about 25% of the world's population. The current coastal urban population of 200 million is projected to almost double in the next 20 to 30 years. This expanding human presence has dramatically changed the coastal natural environment. To meet the growing demand for more housing and other land uses, land has been reclaimed from the sea in coastal areas in many countries, including China, Britain, Korea, Japan, Malaysia, Saudi Arabia, Italy, the Netherlands, and the United States. Coastal areas are often the ultimate discharge zones of regional ground water flow systems. The direct impact of land reclamation on coastal engineering, environment and marine ecology is well recognized and widely studied. However, it has not been well recognized that reclamation may change the regional groundwater regime, including groundwater level, interface between seawater and fresh groundwater, and submarine groundwater discharge to the coast.

This book will first review the state of the art of the recent studies on the impact of coastal land reclamation on ground water level and the seawater interface. Steady-state analytical solutions based on Dupuit and Ghyben-Herzberg assumptions have been derived to describe the modification of water level and movement of the interface between fresh groundwater and saltwater in coastal hillside or island situations. These solutions show that land reclamation increases water level in the original aquifer and pushes the saltwater interface to move towards the sea. In the island situation, the water divide moves towards the reclaimed side, and ground water discharge to the sea on both sides of the island increases. After reclamation, the water resource is increased because both recharge and the size of aquifer are increased.

This book will then derive new analytical solutions to estimate groundwater travel time before and after reclamation. Hypothetical examples are used to examine the changes of groundwater travel time in response to land reclamation. After reclamation, groundwater flow in the original aquifer tends to be slower and the travel time of the groundwater from any position in the original aquifer to the sea becomes longer for the situation of coastal hillside. For the situation of an island, the water will flow faster on the unreclaimed side, but more slowly on the reclaimed side. The impact of reclamation on groundwater travel time on the reclaimed side is much more significant than that on the unreclaimed side. The degree of the modifications of the groundwater travel time mainly depends on the scale of land reclamation and the hydraulic conductivity of the fill materials.

Chapter 1

INTRODUCTION

Most of the large urban centers are coastal and population in coastal areas is expanding rapidly. This expanding human presence has dramatically changed the coastal natural environment, which has made human as the most active geological agent in coastal areas. Coastal areas are faced with a challenge of lack of land due to increase in ecological, economic and social activities, which have led to an increase in pressure on coastal areas and in some cases led to the change of the coastal natural environment. Land reclamation, creating of new land in areas that were initially covered with water, has been a common practice to reduce the increasing pressure in coastal areas (Seasholes, 2003; Lumb, 1976; Suzuki, 2003; Stuyfzand, 1995). Coastal areas are usually the ultimate discharge zones of regional groundwater flow systems. Human activities such as large-scale land reclamation may have severe effect on coastal groundwater regimes.

In the early days, land was reclaimed by draining of swampy or seasonally submerged wetlands. In the coastal areas where intertidal zones have already been reclaimed, coastal reclamation usually means that shallow sea is landfilled extensively to create more useful land area. The analytical studies in this paper focus on the latter case. Land reclamation can increase the land for food production, create new investment into a city, and create new areas for urban development, easing housing shortages. From 1949 to 2001, about 22, 000 km^2 land, which is about half of the total tidal wetlands, along the coast of the Mainland China has been reclaimed. One third of urban area in Hong Kong is reclaimed from the sea. Land reclamation has been also very popular in many other countries including Japan, Korea, Malaysia, USA etc. For example, 45% of Korea's population lives in areas reclaimed from intertidal zone or shallow sea.

While reclamation satisfies the growing needs for land use, it also induces various engineering, environmental and ecological problems such as modification of coastal sediment deposition and erosion, coastal flooding, loss of the coastal habitat and change of marine environment. These problems have been widely discussed (Barnes, 1991; Noske, 1995; Ni et al., 2002; Terawaki et al., 2003). Coastal areas are often the ultimate discharge zones of regional ground water flow systems. Interaction between coastal groundwater and seawater always occurs (Fetter, 1972; Jiao and Tang, 1999), and is expected to be modified by land reclamation. Because the response of a groundwater regime to land reclamation may be slow and not so obvious, this phenomenon has been largely ignored. Stuyfzand (1995) discussed the impact of land reclamation on groundwater quality and future drinking water supply in the Netherlands. Mahamood and Twigg (1995) noted that the water table increased under areas related to reclaimed land in Bahrain Island of Saudi Arabia. A research team in the Department of Earth Sciences, the University of Hong Kong has carried out a research project aiming to understand the impact of large-scale land reclamation on physical and chemical process in coastal groundwater system. They conducted numerical modeling on the impact of land reclamation on groundwater flow and contaminant migration in Penny's Bay and Mid-Level areas, Hong Kong (Jiao, 2000; Jiao, 2002; Jiao et al., 2006). Jiao et al. (2001) presented some analytical solutions on the influence of reclamation on the water level changes in coastal unconfined aquifers. Jiao et al. (2005) hypothesized that there may be various physiochemical reactions between the fill materials and the original marine sediments in a reclamation site and these processes may result in chemicals which may have adverse effect to the coastal plants and organisms. Guo and Jiao (2007) presented some analytical solutions on impact of land reclamation on both the ground water level and the sea water interface with unconfined ground water conditions.

Reclamation is expected to modify the groundwater travel time. Ground water travel time is defined as the time required for a unit volume of ground water to travel between two locations. The travel time is the length of the flow path divided by the velocity, where velocity is the average groundwater flux passing through the cross-sectional area of the geologic medium through which flow occurs, perpendicular to the flow direction, divided by the effective porosity along the flow path. If discrete segments of the flow path have different hydrologic properties the total travel time will be the sum of the travel times for each discrete segment (Code of Federal Regulations, 1988). Groundwater travel times have important implications for the management of water resources, evaluation of the groundwater availability and sustainability, and dating of groundwater. The assessment of groundwater travel times can be used to determine spreading of contaminated plume. The origin of groundwater particles can be tracked if their ages are known. In coastal aquifers, large amounts of dissolved and suspended materials are transported from the land to the sea together with terrestrial groundwater. Discharge of terrestrial groundwater derived from rainfall is a primary component of SGD (submarine groundwater discharge). Thus it is necessary to investigate sensitivity of the groundwater travel time in coastal aquifers.

The ground water transit time is defined as the time interval of water flow from the recharge points to the discharge points (Etcheverry and Perrochet, 2000□Chesnaux et al., 2005; Cornaton and Perrochet, 2005). Groundwater transit times have important applications and their sensitivity has been well investigated. Fórizs and Deák (1998) compared transit times calculated from stable oxygen isotope data and hydraulic modeling. They concluded a time series of groundwater sampling would be necessary when groundwater transit time was calculated from isotope data. Florea and Wicks (2001) constructed laboratory-scale karstic models to investigate transit times by controlling the length of the flow paths, the gradient, the recharge rate, the size of the conduits, and the geomorphic pattern of conduit layout. Their models provided reliable data that could be used in calibrating numerical models. A general theoretical framework to model complete groundwater age and transit time distributions at aquifer scale was presented by Cornaton and Perrochet (2005). The effect of aquifer structure and macro-dispersion on the distributions of age, life expectancy and transit time was discussed by means of analytical and numerical analysis of one- and two-dimensional theoretical flow configurations.

While numerical solutions for determining groundwater age and travel times are available (e.g., Varni and Carrera, 1998), they need more computational efforts and may be applicable for more complicated groundwater system. In many cases, analytical solutions can be used to assess travel times even though their application is limited by a certain number of assumptions. Gelhar and Wilson (1974) developed an analytical solution to calculate the groundwater transit time using the Dupuit assumption. Their analytical solution does not include the influence of the aquifer hydraulic conductivity. Amin and Campana (1996) presented a general lumped parameter mathematical model for estimating mean transit times in hydrologic systems characterized by different mixing regimes. Etcheverry and Perrochet (2000) calculated transit-time distributions based on the reservoir theory without considering the case of surface recharge. More recently, Chesnaux et al. (2005) developed a closed-form analytical solution for calculating groundwater transit times in unconfined aquifers, which included the influence of the hydraulic conductivity of the aquifer. Numerical verification was conducted to testify their analytical solution and comparison was made between their solution and that from Gelhar and Wison (1974). Chesnaux et al. (2005) concluded that the derived closed-form analytical solution was in excellent agreement with the numerical solution. Analytical solutions were also developed for computing the travel time for flow to a pumping well in unconfined aquifer without recharge (Simpson et al., 2003; Chapuis and Chesnaux, 2006).

An important class of ground water flow problems involves steady state, unconfined interface flow in coastal aquifers. Hydrologists have investigated the sea water interface for many decades. While analytical solutions do not directly solve "real-world" problems, they can serve as instructional tools, be used for first-cut engineering analysis and be used as benchmark problems for testing numerical algorithms (Bear et al., 1999). More than a century ago, a relationship resulting from Ghyben (1888) and Herzberg (1901) described that saltwater occurred underground at a depth below sea level of about forty times the freshwater head above the sea level. This relationship, known as Ghyben-Herzberg relation, has been widely used (Bear et al., 1999). With the sharp-interface assumption, many classical analytical solutions have been presented (Henry 1964; Glover 1964; Bear and Dagan 1964; Fetter 1972; Strack 1976). Glover (1964) used the method of complex variables to calculate the flow net and the position of the interface in an infinitely thick confined aquifer. Henry (1964) presented a solution for the position of the interface more precisely with the hodograph method. By comparing this solution with a model obtained under the Dupuit assumption (Fetter, 1994), Henry concluded that the use of the Dupuit assumption could produce sufficiently accurate results for most natural conditions. Fetter (1972) presented a general solution to determine the position of the saline interface under an oceanic island of any shape. A single-potential solution for regional interface problems in coastal aquifers was presented by Strack (1976), which was restricted to cases of steady flow with homogeneous isotropic permeability where the Dupuit assumption was valid.

Both Dupuit and Ghyben-Herzberg assumptions are also used in this paper. It is further assumed that the system is unconfined and all the ground water flow originates as precipitation recharge. The paper concerns only the long-term impact of large-scale land reclamation on coastal groundwater flow system and it is assumed that the ground water flow before reclamation is in a steady state and achieves another steady state after reclamation.

This paper consists of two parts. Based on the recent studies by Jiao et al (2001) and Guo and Jiao (2007), Part one presents the analytical solutions to study the long-term impact of land reclamation on both water level and saltwater interface in a coastal extensive land mass and an island with unconfined ground water conditions. Steady-state analytical solutions have been derived to describe the modification of water level and movement of the saltwater interface in coastal hillside or island situations. The derivation of the analytical solutions was presented in terms of water head. A far more advanced method in terms of the potential function (Strack, 1976) is also presented to obtain the analytical solutions (see appendix A and B). Variable-density flow and solute transport simulations, conducted by the numerical code FEFLOW, are used to evaluate the accuracy of the analytical solutions. A case study based on a major reclamation site in Shenzhen, China is presented to demonstrate the applicability of the analytical solutions.

Part two presents new analytical solutions for the travel time in coastal unconfined aquifers before and after reclamation for cases of both an extensive land mass and an island. Hypothetical examples are utilized to examine the difference of the new analytical solution and the solution developed by Chesnaux et al. (2005) which does not include the influence of the seawater-terrestrial groundwater interface. Sensitivity of travel times to the reclamation scale and hydraulic conductivity of the reclamation material is discussed. With the increase of the concern for pollution of aquifers, it becomes desirable to predict travel paths of contaminated water through the aquifer system. The streamline pattern will give a useful insight into spreading of contaminants. The impact of land reclamation on streamline pattern in the aquifer is investigated. Furthermore, the changes of the flow-velocity distribution in response to land reclamation are discussed.

Chapter 2

IMPACT OF LAND RECLAMATION IN AN EXTENSIVE LAND MASS

Impact of land reclamation on groundwater flow in an extensive land mass was investigated by Jiao et al. (2001) and Guo and Jiao (2007). Problems of combined shallow unconfined interface flow and unconfined flow often occur in coastal aquifers. Figure 1a shows a coastal unconfined ground water system receiving uniform vertical recharge w. The fresh groundwater and the sea water near the coast are separated by a saltwater interface. The hydraulic conductivity of the unconfined aquifer is K_1. The distances from the water divide to the coastline and the tip of the saltwater tongue are represented as L_1 and x_t, respectively (Figure 1a). The head in the fresh water is h above the horizontal impermeable bottom of the aquifer, which is H_0 below sea level. After reclamation (Figure 1b), the coastline moves toward the sea by a distance of L_2, which is hereafter called reclamation length, and the hydraulic conductivity of the fill materials is K_2 (the boundary between K_1 and K_2 is approximated as vertical). The distance from the water divide to the post-reclamation tip is assumed to be x_{tr}. This study focuses primarily on changes of coastal groundwater systems in response to large-scale land reclamation. In this case, it is assumed that the post-reclamation tip of the salt water tongue is located in the reclaimed land when steady-state flow conditions are achieved. The conceptual models in Figure 1 are similar to those presented by Guo and Jiao (2007).

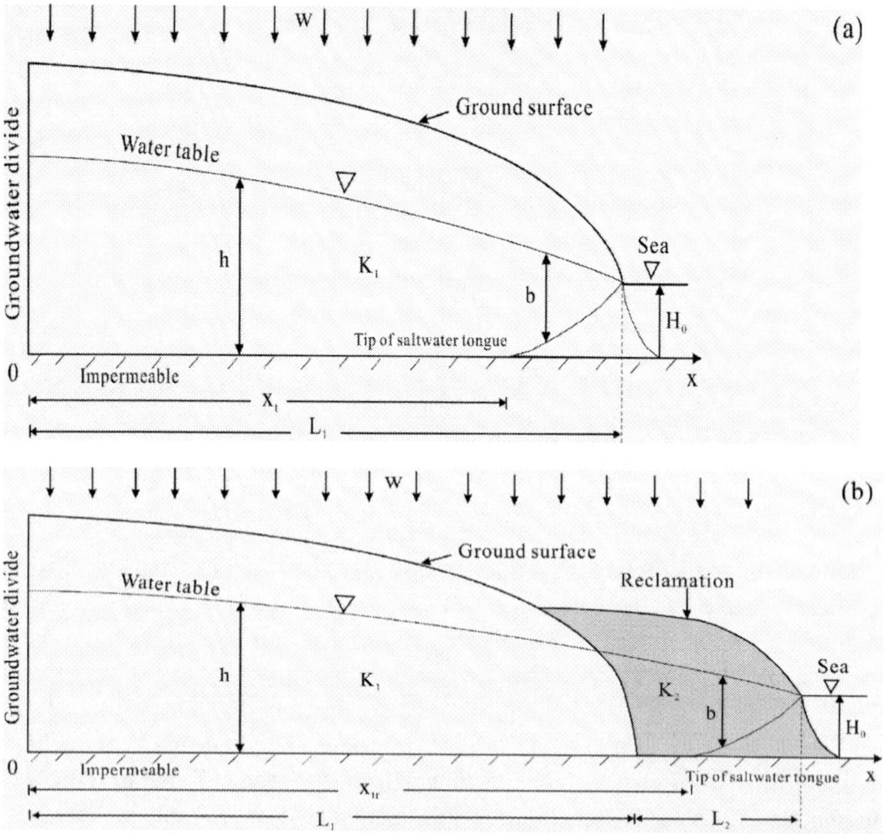

Figure 1. Sketch of an unconfined aquifer system and the seawater-terrestrial groundwater interface in a coastal extensive land mass (a) before reclamation and (b) after reclamation (Guo and Jiao, 2007).

2.1. Impact of Reclamation on Ground Water Level and Saltwater Interface

2.1.1. Solution for Ground Water Level and Saltwater Interface

According to Ghyben-Herzberg assumption (Bear et al. 1999), the seawater and fresh groundwater are assumed to be separated by a sharp interface rather than by a transition zone. The saltwater interface can be regarded as a fixed impermeable boundary at a distance below sea level equal to

$$\overline{z} = \frac{\rho_f (h - H_0)}{\rho_s - \rho_f} \tag{1}$$

where ρ_f and ρ_s are the densities of ground water and seawater. As showed in Figure 1a, the depth of flow in the Dupuit equation, b, is

$$b = h \ (0 \le x \le x_t) \tag{2}$$

$$b = \overline{z} + h - H_0 = \frac{\rho_s (h - H_0)}{\rho_s - \rho_f} \ (x_t \le x \le L_1) \tag{3}$$

The depth of the saltwater interface below sea level $\overline{z}$ is equal to H_0 at $x = x_t$, which leads to

$$h = \frac{\rho_s}{\rho_f} H_0 \qquad (x = x_t) \tag{4}$$

The Dupuit assumption, which states that equipotential surfaces are essentially vertical and the flow horizontal, is a most powerful tool for treating unconfined or unconfined interface flow. Ground water flow in the unconfined aquifer is assumed to satisfy Dupuit assumption and can be described as:

$$\frac{d}{dx}(Kb\frac{dh}{dx}) + w = 0 \tag{5}$$

where b is as defined by expressions (2) and (3), and K is the hydraulic conductivity of the aquifer. Guo and Jiao (2007) presented analytical solutions for the water level and the positions of the saltwater interface before and after reclamation. The main solutions are as follows:

(I) before reclamation

$$h = \sqrt{\frac{w}{K_1}(L_1^2 - x^2) + \frac{\rho_s}{\rho_f} H_0^2} \qquad (0 \le x \le x_t) \tag{6}$$

$$h = \sqrt{\frac{w(\rho_s - \rho_f)}{K_1 \rho_s}(L_1^2 - x^2) + H_0} \qquad (x_t \le x \le L_1) \qquad (7)$$

$$x_t = \sqrt{L_1^2 - \frac{K_1(\rho_s^2 - \rho_s \rho_f)}{w \rho_f^2} H_0^2} \qquad (8)$$

(II) after reclamation

$$h = \sqrt{w\left[\frac{1}{K_1}(L_1^2 - x^2) + \frac{1}{K_2}(2L_1 + L_2)L_2\right] + \frac{\rho_s}{\rho_f} H_0^2} \qquad (0 \le x \le L_1) \qquad (9)$$

$$h = \sqrt{\frac{w}{K_2}\left[(L_1 + L_2)^2 - x^2\right] + \frac{\rho_s}{\rho_f} H_0^2} \qquad (L_1 \le x \le x_{tr}) \qquad (10)$$

$$h = \sqrt{\frac{w(\rho_s - \rho_f)}{K_2 \rho_s}\left[(L_1 + L_2)^2 - x^2\right] + H_0} \qquad (x_{tr} \le x \le L_1 + \qquad (11)$$

$$x_{tr} = \sqrt{(L_1 + L_2)^2 - \frac{K_2(\rho_s^2 - \rho_s \rho_f)}{w \rho_f^2} H_0^2} \qquad (12)$$

2.1.2. Numerical Verification and Comparison

In reality, the seawater-terrestrial groundwater interface is a transition zone so that numerical experiments with a variably-density flow model need to be conducted to check the validity of the derived analytical solutions. The analytical solutions for the water table and the seawater-freshwater interface are here tested numerically for a field-scale problem using the FEFLOW code (Diersch, 2002).

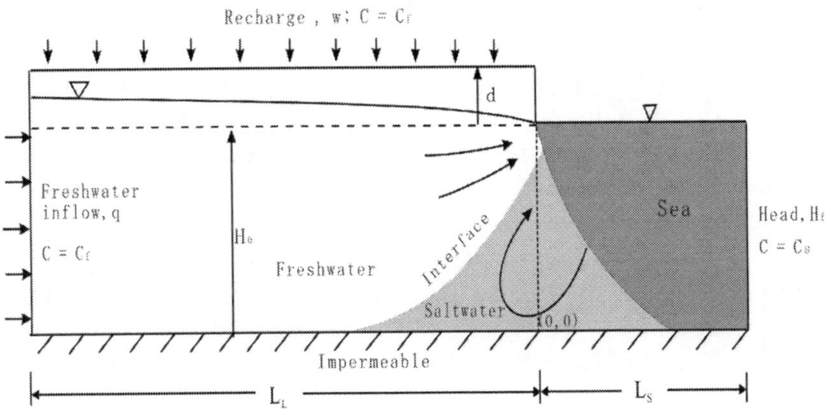

Figure 2. Schematic diagram of the geometry and boundary conditions used in the numerical verification.

Geometry, Aquifer Properties, Boundary Conditions, and Numerical Strategy

Figure 2 shows the geometry and boundary conditions that are considered to verify the previous analytical solutions. It represents a vertical section through a typical coastal unconfined aquifer. The notations L_L and L_S are the landward and seaward lengths. L_L and L_S were determined experimentally so that they didn't influence the shape of the mixing zone significantly (Smith, 2004). The notation d is the distance from the sea level to the land surface. The sea level is assumed to be H_0 above the impermeable bottom boundary. The coordinate origin is set to be on the bottom boundary with the same x coordinate of the shoreline (see Figure 2).

The base of the aquifer at $z = 0$ is assumed to be a no-flow boundary. Neuman boundary conditions are specified along the left model boundary and the landward portion of the top boundary, which respectively represent the terrestrial groundwater inflow and the local recharge. The boundary conditions can be described as follows:

$$C(-L_L, z) = C_f \qquad 0 \le z \le H_0 \tag{13}$$

$$Q(-L_L, z) = w(L_1 + L_2 - L_L) \quad 0 \le z \le H_0 \tag{14}$$

$$C(x, H_0 + d) = C_f \qquad -L_L \leq x \leq 0 \tag{15}$$

$$N(x, H_0 + d) = w \qquad -L_L \leq x \leq 0 \tag{16}$$

where C_f is the salt concentration of the fresh groundwater, Q is the discharge per unit width (L^2/T), N is the uniform recharge across land portion of the upper boundary (L/T), and w is the infiltration rate used in the numerical solution (L/T). L_1 and L_2 are the distance from the water divide to the original coastline and the reclamation length, respectively (Figure 1).

The seabed is assumed to be horizontal in all simulations and the sea level static so that influences of the tidal fluctuations, waves, and the seasonal water exchange between the freshwater and the seawater are neglected. The simplification of the surface water body to be non-penetrating has been adopted in many works (Townley and Davidson, 1988; Nield et al.,1994; Smith and Turner, 2001; Smith, 2004). The boundary conditions along the seabed portion of the top boundary and the right boundary can be described as follows:

$$H_f(x, H_0) = 0 \qquad 0 \leq x \leq L_S \tag{17}$$

$$C(x, H_0) = C_S \qquad 0 \leq x \leq L_S \tag{18}$$

$$H_f(L_S, z) = z + \frac{\rho_s}{\rho_f}(H_0 - z) \qquad 0 \leq z \leq H_0 \tag{19}$$

$$C(L_S, z) = C_S \qquad 0 \leq z \leq H_0 \tag{20}$$

where H_f is the hydraulic head (L) expressed at the fresh groundwater density ρ_f (M/L^3), ρ_s is the density of sea water, C is the salt concentration (M/L^3), and C_S is the salt concentration of the seawater. The so-called "boundary constraint" condition available in FEFLOW simulator is utilized to achieve the described mass boundary.

Three hypothetical cases were simulated to verify the accuracy of the analytical solutions (Table 1). An iterative solution for steady flow and mass transport was obtained for case 1. However, the solutions could not converge for cases 2 and 3, then transient model runs were performed to achieve quasi-steady states.

The simulation parameters are presented in Table 2. These parameters are based on those used by Smith (2004) for similar coastal groundwater problems.

The numerical solution strategy adopted in the simulation is listed in Table 3. The Boussinesq approximation is applied so that density variations are included by the buoyancy term of the Darcy equation only. This approximation can also simplify the mathematical model of the variable density flow. In addition, the so-called no-upwinding approach is used to keep numerical dispersion as low as possible (Kolditz et al., 1998). Furthermore, the influence of the viscosity on the hydraulic conductivity is neglected in this study.

Table 1. FEFLOW Model Inputs for the Hypothetical Simulation Cases

Case	K_1 *m/day*	K_2 *m/day*	L_L *m*	L_S *m*	*w* *m/day*	L_1 *m*	L_2 *m*	H_0 *m*	*Description*
Case 1	3	---	200	50	0.0005	1000	---	20	Before Reclamation
Case 2	3	5	200	50	0.0005	1000	500	20	After Reclamation
Case 3	3	10	400	100	0.0005	1000	500	20	After Reclamation

Table 2. Simulation Parameters for the Numerical Verification

Symbol	*Quantity*	*Value*	*Unit*
D_m	coefficient of molecular diffusion	8.64×10^{-5}	m^2/day
φ	Porosity	0.3	---
C_f	salt concentration of fresh groundwater	0	mg/L
C_S	salt concentration of seawater	35,000	mg/L
$\bar{\alpha}$	density difference ratio	0.025	---
α_L	longitudinal dispersivity of solute	5	m
α_T	transverse dispersivity of solute	0.5	m

Comparison of the results from numerical simulation and those from the analytical solution

Three hypothetical cases (Table 1) are utilized to verify the accuracy of the analytical solutions. Figure 3 depicts white streamlines in the mixing zone superimposed on shaded concentration fringes, which represent the density-driven seawater circulation in the coastal aquifer. This kind of circulation is induced by the displacement of the freshwater by the seawater and the salt dispersion in the mixing zone. A discharge zone forms near the coastline (Figure 3), inducing a natural, realistic shape of the transition zone. It can be inferred from Equation (12) that the post-reclamation seawater intrusion length, $L_1+L_2-x_{tr}$, increases with the hydraulic conductivity of the reclamation material, K_2. The transition zone in Figure 3(c) tends to be more landward than that in Figure 3(b) because the hydraulic conductivity of the aquifer is increased from 5m/day to 10 m/day.

Table 3. Numerical Specifications in FEFLOW

Topic	***Quantity***
Mesh type	3-noded triangle elements (82,162 to 555,696)
Finite element scheme	no upwinding (best-accurate Galerkin-based formulation)
Problem class	combined flow and mass transport
Time stepping scheme	automatic time step control via predictor-corrector schemes (AB/TR time integration scheme)
Mass transport equation	convective form transport
Iterative solver	preconditioned conjugate gradient PCG
Density coupling	Boussinesq approximation
Evaluation of element integral	standard Gauss quadrature
Velocity approximation	improved consistent velocity approximation (by Frolkovic-Knabner)

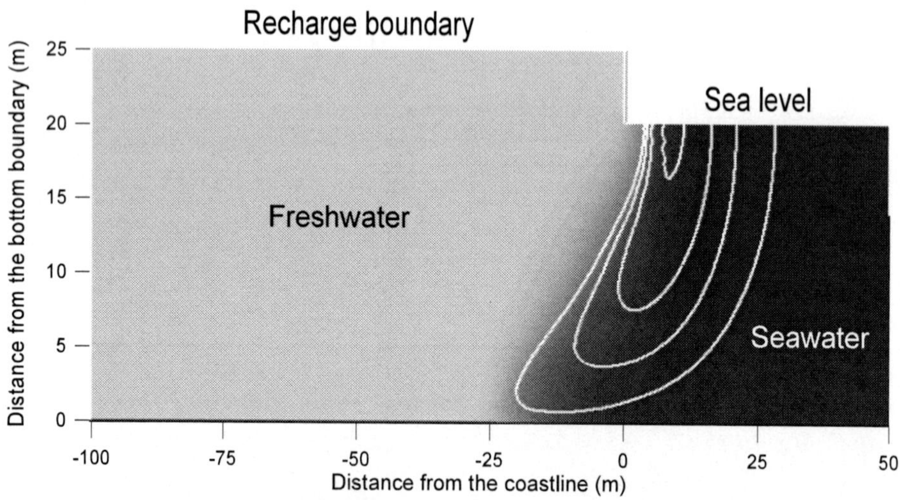

(a)

Figure 3. (Continued on next page.)

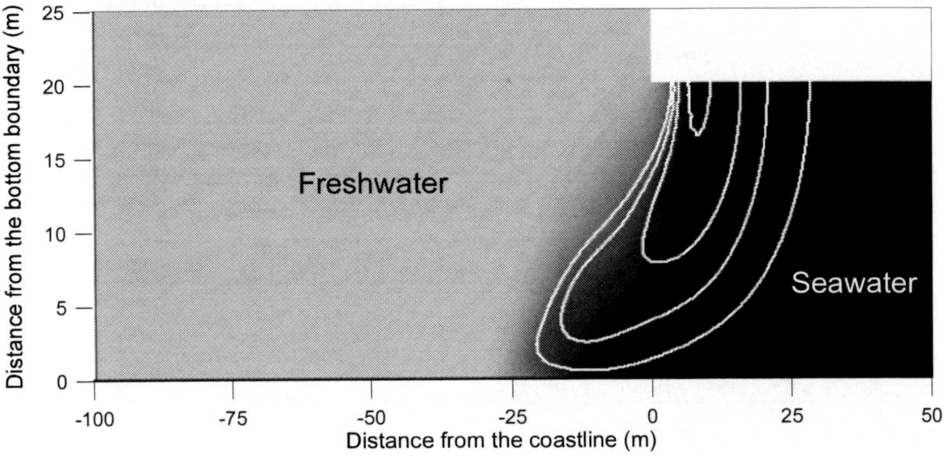

(b)

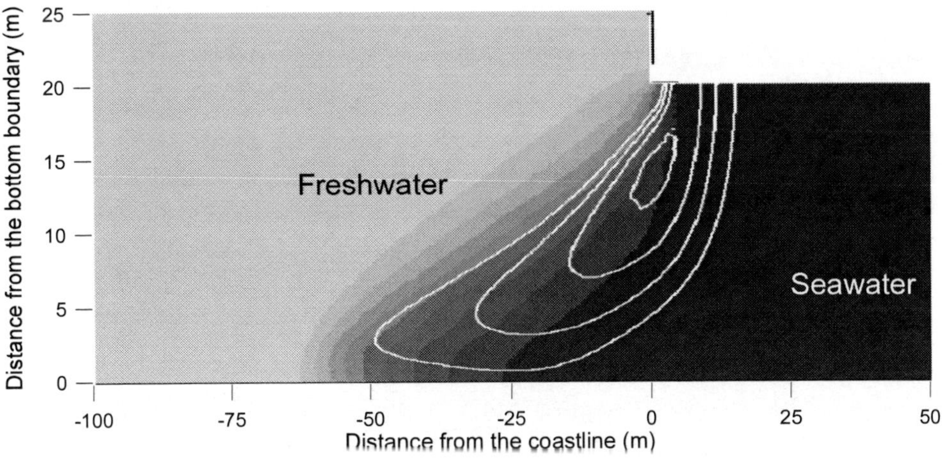

(c)

Figure 3. Steady-state simulation results for the hypothetical examples. White streamlines superimposed on shaded concentration fringes show the seawater circulations. Only results of a horizontal range of 150m besides the coastline are presented in order to compare the simulated mixing zones in different cases. (a) Hydraulic conductivity of the aquifer is 3m/day and no reclamation occurs. (b) Hydraulic conductivities of the original aquifer and the reclaimed aquifer are 3m/day and 5m/day, respectively. (c) Hydraulic conductivities of the original aquifer and the reclaimed aquifer are 3m/day and 10m/day, respectively.

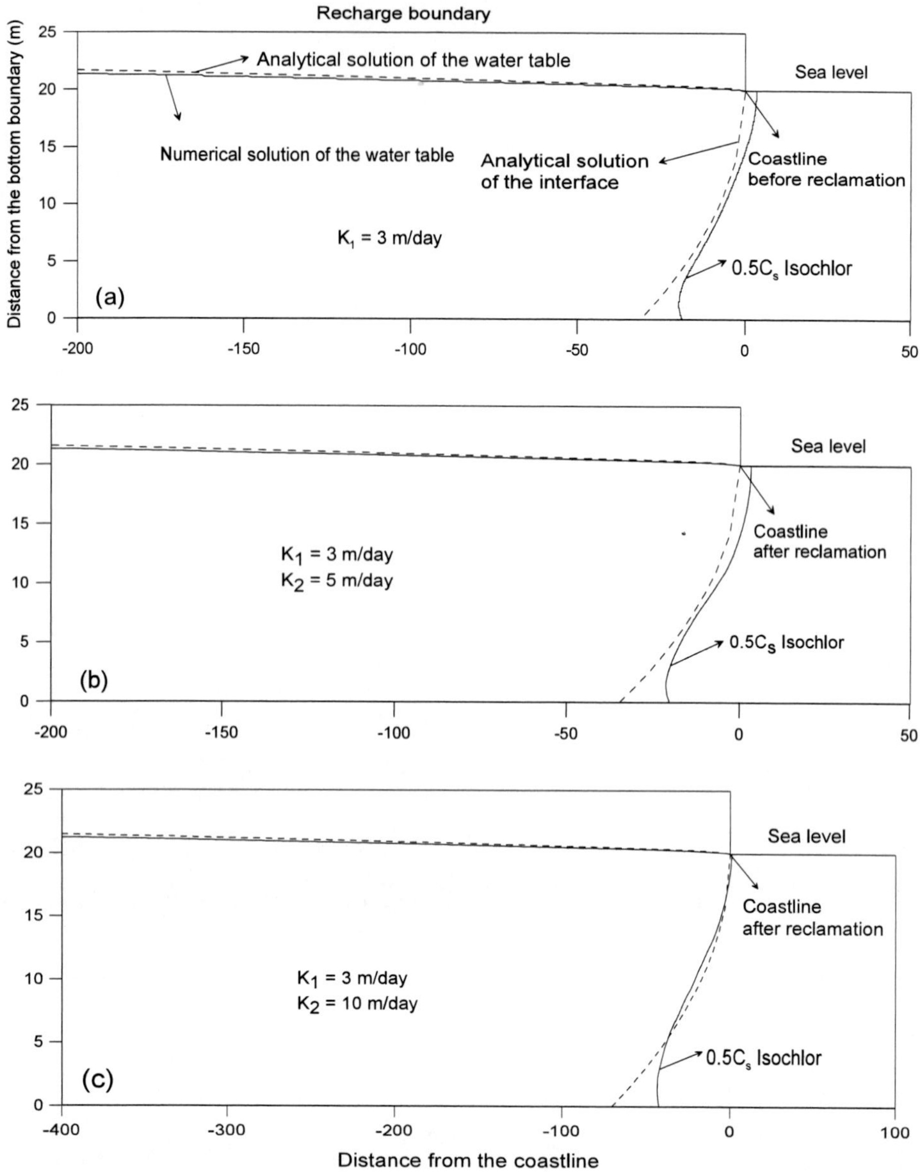

Figure 4. Steady state $0.5C_S$ isochlors compared with saltwater interfaces obtained from the analytical solutions for three different cases. (a) Hydraulic conductivity of the aquifer is 3m/day and no reclamation occurs. (b) Hydraulic conductivities of the original aquifer and the reclaimed aquifer are 3m/day and 5m/day. (c) Hydraulic conductivities of the original aquifer and the reclaimed aquifer are 3m/day and 10m/day.

Figure 4 compares the 0.5Cs isochlors at steady state for the three cases in Table 1 with the corresponding saltwater interfaces obtained from the analytical solutions. The accuracy of the analytical solution is somewhat limited by assumptions such as Dupuit-type flow and the Ghyben-Herzberg relation. Because of these assumptions, the saltwater interfaces from the analytical solutions do not completely agree with the 0.5Cs isochlors produced by the density-driven transport model. The 0.5 Cs isochlors are a little further seaward than the sharp interface obtained from the analytical solutions. Comparisons of the water tables between the numerical solutions and the analytical solutions show that the water tables calculated from the analytical solutions are slightly higher than those from the numerical solutions. However, these differences are not very significant and the maximal difference is 0.5m, which is only 2.5% of the elevation of the sea level in relation to the bottom boundary. Thus the simplifying assumptions used in the analytical studies are testified to be reasonable.

2.1.3. Discussion of the Analytical Solutions

Change of the water level in the original aquifer

The change of the water table in the original aquifer between $x = 0$ and $x = L_1$ induced by reclamation can be calculated from Equations (6), (7) and (9), which yields

$$\Delta h = \sqrt{w\left[\frac{1}{K_1}(L_1^2 - x^2) + \frac{1}{K_2}(2L_1 + L_2)L_2\right] + \frac{\rho_s}{\rho_f}H_0^2} - \sqrt{\frac{w}{K_1}(L_1^2 - x^2) + \frac{\rho_s}{\rho_f}H_0^2}$$

$$(0 \le x \le x_t) \qquad (21)$$

$$\Delta h = \sqrt{w\left[\frac{1}{K_1}(L_1^2 - x^2) + \frac{1}{K_2}(2L_1 + L_2)L_2\right] + \frac{\rho_s}{\rho_f}H_0^2} - \sqrt{\frac{w(\rho_s - \rho_f)}{K_1\rho_s}(L_1^2 - x^2)} - H_0$$

$$(x_t < x \le L_1) \qquad (22)$$

Equations (21) and (22) can be utilized to analyze sensitivity of the ground water table change to the hydraulic conductivity of the reclamation material and the scale of reclamation. The change of the water level at the original coastline, which is the maximum change in the domain, can be readily obtained by setting $x = L_1$ in Equation (22), which leads to

$$\Delta h = \sqrt{\frac{w}{K_2}(2L_1 + L_2)L_2 + \frac{\rho_s}{\rho_f} H_0^2} - H_0 \tag{23}$$

Equation (23) indicates the maximum buildup of the water level at the original coastline is independent of K_1, and increases with L_1, L_2, and the ratio between the recharge rate w and K_2. For a specific coastal area, the parameters L_1 and w are fixed so that the water level buildup at the original coastline mainly depends on the hydraulic conductivity of the reclamation material K_2 and the reclamation length L_2. The reclamation tends to have a significant damming effect when L_1 is great and K_2 is low. This conclusion is the same as that of the study by Jiao et al (2001) which ignored the seawater interface.

Displacement of the tip of the saltwater tongue

Due to reclamation the position of the tip of the saltwater tongue moves towards the sea by a distance of $x_{tr} - x_t$, which can be obtained from Equations (8) and (12):

$$x_{tr} - x_t = \sqrt{(L_1 + L_2)^2 - \frac{k_2(\rho_s^2 - \rho_s\rho_f)}{w\rho_f^2} H_0^2} - \sqrt{L_1^2 - \frac{k_1(\rho_s^2 - \rho_s\rho_f)}{w\rho_f^2} H_0^2} \tag{24}$$

For a particular coastal area, parameters L_1, K_1, w, and H_0 are fixed, the change of the tip of the saltwater tongue depends mainly on L_2 and K_2. As mentioned early, this study assumes that the tip of the saltwater tongue after reclamation is located in the reclaimed land when steady state flow conditions are reached. In this case, the value $x_{tr} - x_t$ is always positive, i.e. the tip of the saltwater interface will be pushed seaward after reclamation. Equation (24) shows that displacement of the tip, $x_{tr} - x_t$, is significant if the reclamation length L_2 is great and the hydraulic conductivity of the reclamation material K_2 is low.

2.1.4. Hypothetical Example

The impact of the reclamation on the ground water level and the seawater-freshwater interface is discussed here with a hypothetical example. Assume that the saturated hydraulic conductivity K_1 of the aquifer is 0.1 m/day and the distance from the ground water divide to the original coastline, L_1, is 1000m. The densities of seawater and freshwater are 1.025 g/cm^3 and 1.000 g/cm^3, respectively. The infiltration rate is 0.0005 m/day and H_0 =20m.

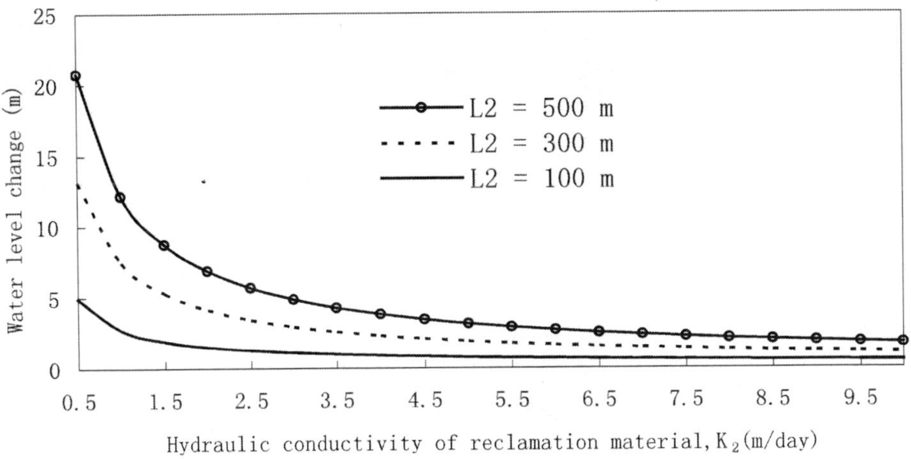

Figure 5. Change of ground water level at the original coastline with hydraulic conductivity of the reclamation materials when K_1=0.1m/day, L_1=1000m, ρ_s =1.025g/cm^3, ρ_f =1.000g/cm^3, w=0.0005m/day, and H_0=20m.

The land reclamation will increase the water level in the original aquifer and push the saltwater interface seaward. A comparison of predicted water levels at the original coastline as a function of hydraulic conductivity and reclamation length shows that the water level rise at the original coastline after reclamation decreases with K_2 and increases with L_2 (Figure 5). The buildup of the water level at the original coastline can be great and is sensitive to K_2 when K_2 is low. When K_2 is greater than 5 m/day, the water-level rise tends to be small and insensitive to K_2.

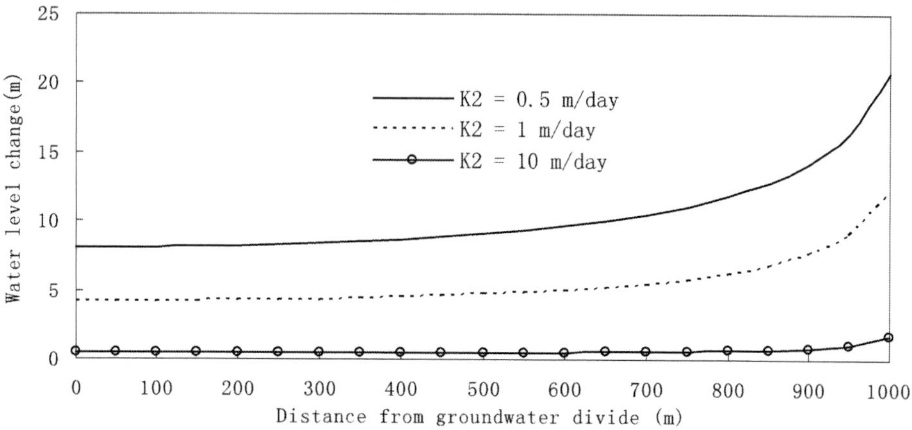

Figure 6. Change of ground water level in the original unconfined aquifer with distance from the ground water divide when K_1=0.1m/day, L_1=1000m, L_2=500m, ρ_s =1.025g/cm^3, ρ_f =1.000g/cm^3, w=0.0005m/day, and H_0=20m.

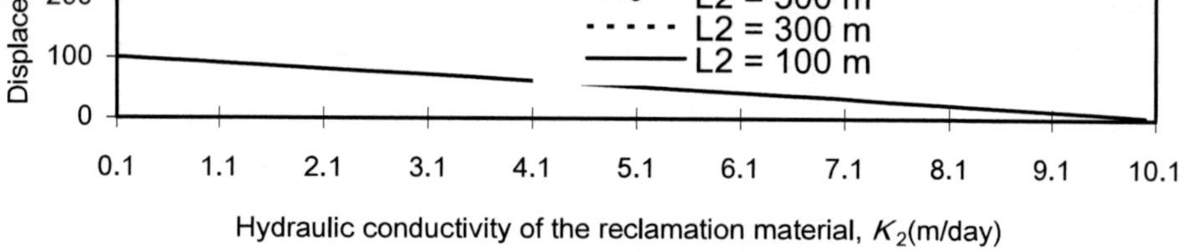

Figure 7. Displacement of the tip of the saltwater tongue with hydraulic conductivity of the reclamation materials for different reclamation lengths when K_1=0.1m/day, L_1=1000m, ρ_s =1.025g/cm^3, ρ_f =1.000g/cm^3, w=0.0005m/day, and H_0=20m (Guo and Jiao, 2007).

Figure 6 shows the change of ground water level in the original unconfined aquifer with distance from the ground water divide when L_2=500m. The rise of the water level increases with the distance from the ground water divide and decreases with hydraulic conductivity of the fill materials. For any fixed value of K_2, the water level change is sensitive to distance from the water divide and can be significant at positions close to the original coastline. The water-level rise is much more sensitive to K_2, and the maximum rise is located at the original coastline. With $L_2 = 500$ m the water level rise at the original coastline is 20.7 m for $K_2 = 0.5$ m/day and only 1.7 m for $K_2 = 10$ m/day.

Figure 7 shows the displacement of the tip of the saltwater tongue, which is defined as x_{tr}-x_t, as a function of L_2 and K_2. The displacement of the tip decreases with K_2 and increases with L_2, and is usually less than L_2 when $K_2 > K_1$. The displacement is slightly greater than L_2 when $K_2 = K_1$ because the tip is pushed much more seaward due to the increased recharge in the reclaimed land. The displacement of the tip decreases almost linearly with K_2, indicating that the post-reclamation saltwater tongue can be significantly longer than the initial one when K_2 is great. After reclamation, the saltwater interface is pushed seaward, which may benefit groundwater wells screened close to the saltwater interface.

2.1.5. Application: a Case Study in Shenzhen, China

The study site covers a major reclamation area to the southwest of Shenzhen, China (see Figure 8). The rapid urban development has led to a sharp increase in the demand for usable land in this area during the past 20 years, and large-scale land reclamation by filling the shallow sea has been a common practice to ease this demand. Figure 8 shows the original coastline in 1983 and the coastline in 2000 around the Deep Bay, Shenzhen. The water divide is located about 4.2 km north of the reclamation area and lies along topographical peaks and ridges. Between 1983 and 2000, the coastline has been pushed towards the sea by a distance of approximately 700 m due to a series of reclamation projects.

The study area around Deep Bay is underlain mainly by weathered igneous rock. Ding (2006) analyzed the measured hydraulic conductivity values of the weathered granite and concluded that average hydraulic conductivity value was of the order of 10^{-6} m/s. In the analytical study, the hydraulic conductivity of the original aquifer K_1 is assumed to be 1.0×10^{-6} m/s. The approximate hydraulic conductivity of the fill material in this area is of the order of 10^{-4} m/s. The climate of this area is subtropical humid with hot wet summer and mild dry winter, and the average annual rainfall is about 1837 mm. The recharge coefficient is here taken as 0.1 over a long-term basis, which leads to an annual infiltration rate of about 184mm. The average reclamation thickness in Deep Bay is about 10 m. According to Equations (22) and (24), the estimated values are very sensitive to hydraulic conductivity of the fill material K_2, which is uncertain in the field. If $K_2 = 2.0 \times 10^{-4}$ m/s, the estimated change of the water level at the original coastline is 6.9m. Because of the reclamation, the saltwater tongue will be pushed to the new coastline, and the displacement of the tip of the tongue is estimated to be 691m based on (24). The change of the water level at the original coastline and the displacement of the tip are estimated to be 3.3m and 677m when K_2 is increased to be 5.0×10^{-4} m/s.

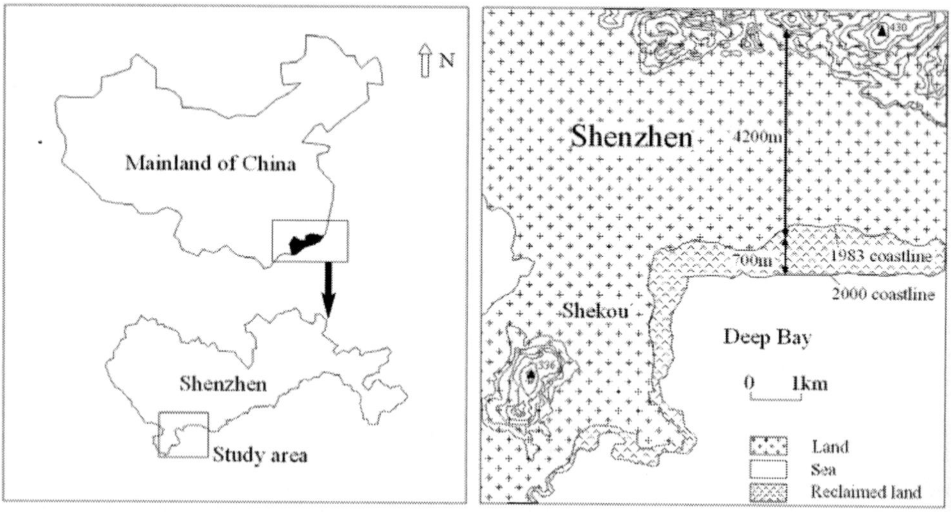

Figure 8. Change of the coastline due to reclamation in Shenzhen, China between year 1983 and year 2000.

2.2. Impact of Reclamation on Ground Water Travel Time

According to the Dupuit theory, the flow in the unconfined aquifer is considered horizontal and one dimensional, and the ground water discharge per unit width of aquifer is defined as:

$$Q_x = -Kb\frac{dh}{dx} = q_x b \tag{25}$$

where Q_x is the ground water discharge per unit width (L^2/T), K is the saturated hydraulic conductivity(L/T), b is the thickness of the fresh groundwater (L) , h is the hydraulic head (L), and q_x is specific discharge or Darcy flux (L/T).

In the unconfined systems (Figure 1), infiltration occurs at a constant rate w along the entire upper boundary of the aquifer. The ground water discharge per unit width of aquifer affected by a uniform recharge is

$$Q_x = w\,x \tag{26}$$

Since the vertical component of the flow velocity is neglected, the ground water velocity v (L/T) can be expressed as:

$$v_x = \frac{dx}{dt} = \frac{q_x}{n_e} \tag{27}$$

where n_e is the effective porosity of the porous medium. Closed-form solutions for travel time t (T) of a water particle to move between two arbitrary positions x_i (at time t_i) and x (at time t) in the unconfined aquifer can be obtained by combining Equations (25), (26) and (27), which leads to

$$\int_{t_i}^{t} dt = \frac{n_e}{w}\int_{x_i}^{x} \frac{b(\tau)}{\tau}\,d\tau \tag{28}$$

Based on Equation (28), Chesnaux et al. (2005) developed an analytical solution to calculate ground water transit time in a Dupuit-type flow system bounded by lakes or rivers (Figure 9). Figure 9 shows the unconfined system used for developing the analytical solution. The left-hand boundary is groundwater divide, and ground water discharges through the right-hand fixed-head boundary. The solution of the transit time from an arbitrary point at which $x = x_i$ (assuming the travel path begins at the water table) can be expressed as:

$$t(x_i) = n_e\sqrt{\frac{\zeta}{wK}}\left(L'\sqrt{\frac{1}{L'^2}-\frac{1}{\zeta}} - x_i\sqrt{\frac{1}{x_i^2}-\frac{1}{\zeta}} + \ln\frac{\frac{\sqrt{\zeta}}{x_i}+\sqrt{\frac{\zeta}{x_i^2}-1}}{\frac{\sqrt{\zeta}}{L'}+\sqrt{\frac{\zeta}{L'^2}-1}}\right) \tag{29}$$

where $\zeta = L'^2 + Kh_{L'}^2 / w$, L' represents the length of the unconfined aquifer system, and $h_{L'}$ is the fixed-head of the right-hand boundary.

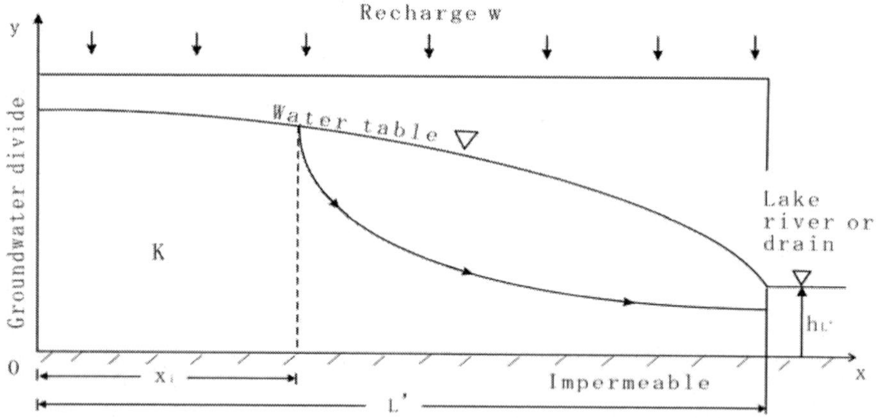

Figure 9. The unconfined system used for developing the analytical solution by Chesnaux et al. (2005). The left-hand boundary is groundwater divide, and ground water discharges through the right-hand fixed-head boundary.

2.2.1. Analytical Solution for the Ground Water Travel Time before Reclamation

Analytical solution

When seawater-freshwater interface is considered, the thickness of the cross-sectional area through which flow occurs will not be always equal to the water head. Change of thickness of freshwater with distance towards the left boundary can be obtained from Equations (2), (3), (6) and (7). The final solutions are

$$b = h = \sqrt{\frac{w}{K_1}(L_1^2 - x^2) + \frac{\rho_s}{\rho_f} H_0^2} \qquad (0 \le x \le x_t) \qquad (30)$$

$$b = \frac{\rho_s (h - H_0)}{\rho_s - \rho_f} = \sqrt{\frac{w\rho_s}{K_1(\rho_s - \rho_f)}(L_1^2 - x^2)} \qquad (x_t \le x \le L_1) \qquad (31)$$

Substituting Equations (30) and (31) into Equation (28), one has:

$$\int_{t_i}^{t} dt = \frac{n_e}{\sqrt{wK_1}} \int_{x_i}^{x} \frac{\sqrt{\omega^2 - x^2}}{x} dx \qquad (0 \le x \le x_t) \qquad (32)$$

$$\int_{t_i}^{t} dt = n_e \sqrt{\frac{\rho_s}{wK_1(\rho_s - \rho_f)}} \int_{x_i}^{x} \frac{\sqrt{L_1^2 - x^2}}{x} dx \qquad (x_t \le x \le L_1) \qquad (33)$$

where

$$\omega = \sqrt{L_1^2 + \frac{\rho_s K_1}{\rho_f w} H_0^2}$$

The solution to Equations (32) and (33) leads to:

$$\int_{t_i}^{t} dt = n_e \sqrt{\frac{1}{wK_1}} \left[\sqrt{\omega^2 - x^2} - \sqrt{\omega^2 - x_i^2} - \omega \ln\left(\frac{\omega + \sqrt{\omega^2 - x^2}}{\omega + \sqrt{\omega^2 - x_i^2}} \frac{x_i}{x} \right) \right] \qquad (0 \le x \le x_t) \qquad (34)$$

$$\int_{t_i}^{t} dt = n_e \sqrt{\frac{\rho_s}{(\rho_s - \rho_f) w K_1}} \left[\sqrt{L_1^2 - x^2} - \sqrt{L_1^2 - x_i^2} - L_1 \ln\left(\frac{L_1 + \sqrt{L_1^2 - x^2}}{L_1 + \sqrt{L_1^2 - x_i^2}} \frac{x_i}{x} \right) \right]$$

$$(x_t \leq x \leq L_1) \tag{35}$$

Integrating Equations (34) and (35) with respect to x from x_0 to x_l yields the travel time of water flow from one position ($x = x_0$) to another position ($x = x_l$) in the unconfined aquifer before reclamation. If $x_l = L_l$, the resultant travel time is the transit time of a particle of water originating at $x = x_0$ before reclamation. The total travel time will be the sum of the travel times for each discrete segment if discrete segments of the flow path have different aquifer properties.

Comparison of the solutions with and without considering seawater interface

A hypothetical example is used here to compare the new analytical solution with the transit time solution developed by Chesnaux et al. (2005), which didn't include the influence of seawater-freshwater interface and can be expressed as Equation (29). Assume the distance from the ground water divide to the coastline is 1000m, and the effective porosity of the aquifer is 0.3. The difference between the groundwater travel time with and without the seawater-freshwater interface is defined as

$$\Delta t = t_{New} - t_{Chesnaux} \tag{36}$$

where t_{New} and $t_{Chesnaux}$ represent the groundwater travel time calculated from the new analytical solution in this paper and from Equation (29) developed by Chesnaux et al. (2005), respectively.

Figure 10 shows the differences (defined by Equation 36) of transit times to the sea from different positions in the aquifer with distance from the water divide for different sea levels when w = 0.0005 m/day and K_l = 3 m/day. The calculated water level becomes higher when the impact of the seawater-freshwater interface is considered. Thus, the thickness of the freshwater between the water divide and the tip of the saltwater tongue will be greater, thereby decreasing the flow velocity. However, the groundwater flow velocity tends to be greater in the area between the tip and the coastline due to decrease of the freshwater thickness in this area. Therefore, as shown in Figure 10, differences of transit times decrease with the distance from groundwater divide, and tend to be positive close to the water divide and negative near the coastline. When H_0 is greater, differences of the transit times will be more significant because the seawater intrusion length tends to be greater according to Equation (8). With H_0 = 30 m the difference is 188.1 days for x = 50m and -424.8 days for x=950m.

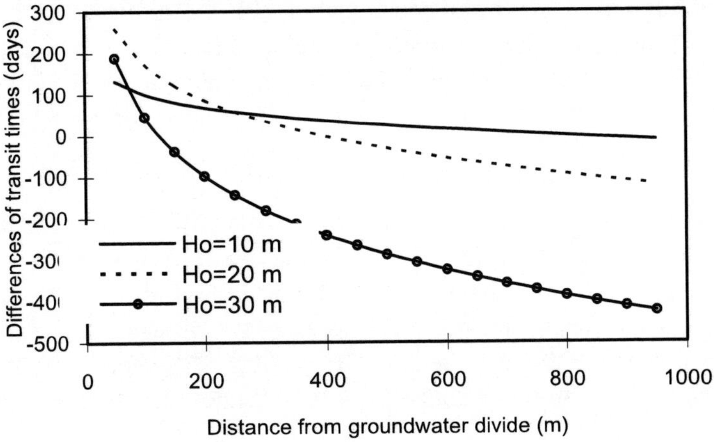

Figure 10. Differences of transit times to the sea from different positions in the aquifer with distance from the water divide for different sea levels when K_l=3m/day, L_l=1000m, ρ_s =1.025g/cm^3, ρ_f =1.000g/cm^3, and w=0.0005m/day.

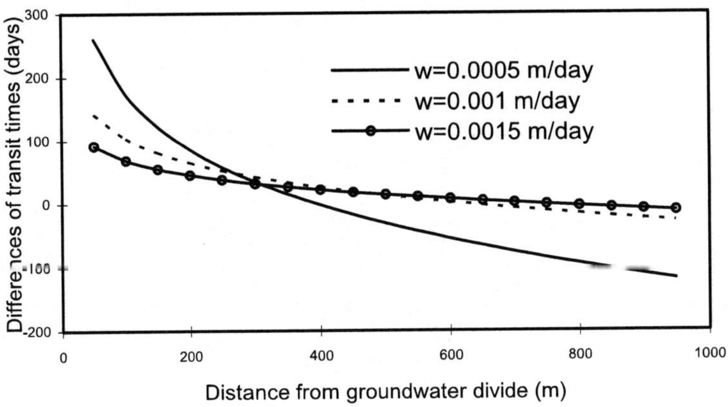

Figure 11. Differences of the transit times to the sea from different positions in the aquifer with distance from the left water divide for different infiltration rates when K_l=3m/day, L_l=1000m, ρ_s =1.025g/cm^3, ρ_f =1.000g/cm^3, and H_0 =20m.

Figure 11 shows differences of transit times to the sea from different positions in the aquifer with distance from the water divide for different infiltration rates when K_l = 3 m/day and H_0 =20m. The curves in Figure 11 show the similar trends as those in Figure 10. According to Equation (8), the influence of the seawater intrusion will be great when the recharge rate *w* is low. Therefore, differences of the transit times to the sea tend to be more significant when the recharge rate *w* is low. With *w* = 0.0005 m/day the difference is 260.1 days for *x* = 50m and -118.5 days for *x*=950m.

Figure 12 shows differences of the transit times with distance from the water divide for different hydraulic conductivities of the aquifer when *w* = 0.0005 m/day and H_0 =20m. According to Equation (8), the seawater intrusion length, L_l-x_t , becomes longer when K_l is increased. In this case, the flow-velocity increase due to decrease of the freshwater thickness between the tip and the coastline will be more significant and be in a greater area. Thus, differences of transit times decrease with K_l. With the distance from the groundwater divide increasing, the impact of the flow-velocity increase close to the coastline on the transit time will become stronger. At the same time, the influence of the flow-velocity decrease due to rise of the water table away from the coastline will diminish. Therefore, differences of the transit times decrease with the distance from the groundwater divide.

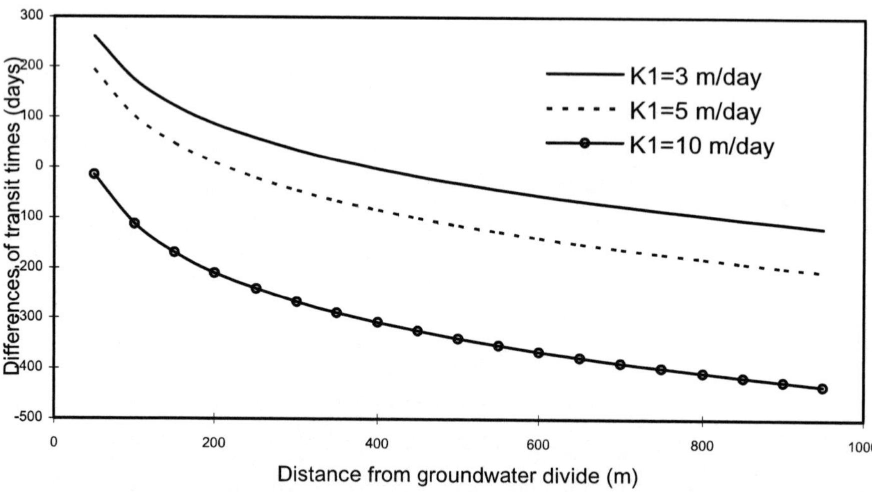

Figure 12. Differences of the transit times to the sea from different positions in the aquifer with distance from the water divide for different hydraulic conductivities of the aquifer when L_l=1000m, ρ_s =1.025g/cm^3, ρ_f =1.000g/cm^3, *w* = 0.0005 m/day, and H_0 =20m.

2.2.2. Analytical Solution for the Ground Water Travel Time after Reclamation

Analytical solution

After reclamation the coastline is pushed seaward by a distance of L_2 so that an additional aquifer forms and rain recharge takes place over a larger area (Figure 1b). The change of the freshwater thickness with distance from the left boundary can be obtained from Equations (9), (10) and (11) as follows:

$$b = h = \sqrt{w\left[\frac{1}{K_1}(L_1^2 - x^2) + \frac{1}{K_2}(2L_1 + L_2)L_2\right] + \frac{\rho_s}{\rho_f}H_0^2} \qquad (0 \le x \le L_1) \tag{37}$$

$$b = h = \sqrt{\frac{w}{K_2}\left[(L_1 + L_2)^2 - x^2\right] + \frac{\rho_s}{\rho_f}H_0^2} \qquad (L_1 \le x \le x_{tr}) \tag{38}$$

$$b = \frac{\rho_s(h - H_0)}{\rho_s - \rho_f} = \sqrt{\frac{w\rho_s}{K_2(\rho_s - \rho_f)}\left[(L_1 + L_2)^2 - x^2\right]} \qquad (x_{tr} \le x \le L_1 + L_2) \tag{39}$$

Equations (37), (38) and (39) can be used to describe relation between thickness of the vertical cross section through which water flow occurs and the distance towards the water divide. Substituting these equations to equation (28) respectively and integrating the derived equations, the following equations are obtained:

$$\int_{t_i}^{t} dt = n_e\sqrt{\frac{1}{wK_1}}\left[\sqrt{\varphi^2 - x^2} - \sqrt{\varphi^2 - x_i^2} - \varphi \ln\left(\frac{\varphi + \sqrt{\varphi^2 - x^2}}{\varphi + \sqrt{\varphi^2 - x_i^2}}\frac{x_i}{x}\right)\right]$$

$$(0 \le x \le L_1) \tag{40}$$

$$\int_{t_i}^{t} dt = n_e \sqrt{\frac{1}{wK_2}} \left[\sqrt{\lambda^2 - x^2} - \sqrt{\lambda^2 - x_i^2} - \lambda \ln \left(\frac{\lambda + \sqrt{\lambda^2 - x^2}}{\lambda + \sqrt{\lambda^2 - x_i^2}} \frac{x_i}{x} \right) \right]$$

$$(L_1 \leq x \leq x_{tr}) \qquad (41)$$

$$\int_{t_i}^{t} dt = n_e \sqrt{\frac{\rho_s}{(\rho_s - \rho_f) w K_2}} \left[\sqrt{L^2 - x^2} - \sqrt{L^2 - x_i^2} - L \ln \left(\frac{L + \sqrt{L^2 - x^2}}{L + \sqrt{L^2 - x_i^2}} \frac{x_i}{x} \right) \right.$$

$$(x_{tr} \leq x \leq L_1 + L_2) \qquad (42)$$

where

$$\varphi = \sqrt{L_1^2 + \frac{K_1}{K_2}(2L_1 + L_2)L_2 + \frac{\rho_s K_1 H_0^2}{\rho_f w}}, \; \lambda = \sqrt{(L_1 + L_2)^2 + \frac{\rho_s K_2 H_0^2}{\rho_f w}}.$$

Discussion of the analytical solutions

The changes of groundwater travel times due to the coastal reclamation are here discussed with a hypothetical example. Assume that the saturated hydraulic conductivity of the original aquifer is 3 m/day and the distance from the ground water divide to the original coastline is 1000m. The infiltration rate is 0.0005 m/day and H_0 =20m.

Figure 13 shows changes of the transit times from a point at which x = 800m to the sea with hydraulic conductivity of the reclamation materials for different reclamation lengths. The increase of the transit times due to reclamation decreases when K_2 increases and L_2 decreases. This is easily understood because the damping effect due to reclamation will be more significant when K_2 decreases and L_2 increases. With K_2 = 0.5 m/day the change is 29.7 years for L_2 = 500 m and 5.5 years for L_2 = 100 m, showing that changes of transit times are very sensitive to the reclamation length. Changes of the transit times tend to be low and insensitive to K_2 when K_2 is greater than 3m/day.

Figure 14 shows how transit times to the sea from different positions in the original aquifer change with the distance from the water divide for different hydraulic conductivity of the reclamation materials when L_2=500m. After reclamation, the water table in the whole original aquifer will increase so that the thickness of freshwater is increased and the water will flow more slowly.

Therefore, as showed in Figure 14, the transit time changes increase from the original coastline to the water divide and tend to be more significant when K_2 becomes low. With $K_2 = 1$ m/day the change is 47.7 years for $x = 200$ m and 19.6 years for $x = 950$ m.

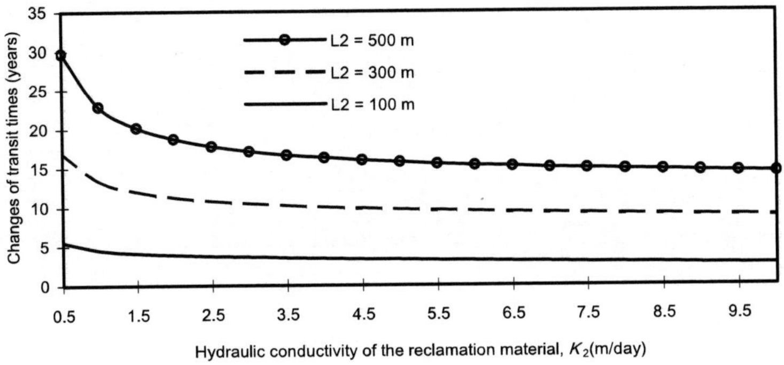

Figure 13. Changes of transit times from a point at which x = 800m to the sea with hydraulic conductivity of the reclamation material for different reclamation lengths when $K_1 = 3$ m/day, $L_1 = 1000$ m, $\rho_s = 1.025$ g/cm^3, $\rho_f = 1.000$ g/cm^3, $w = 0.0005$ m/day, and $H_0 = 20$m.

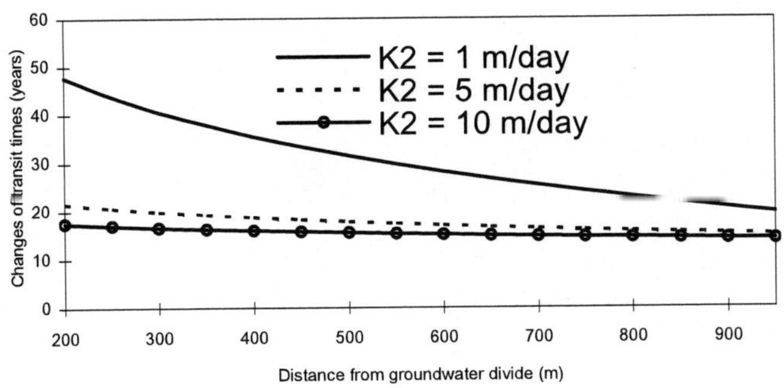

Figure 14. Changes of transit times to the sea from different positions in the original aquifer with distance from the water divide for different hydraulic conductivity of the reclamation material when $K_1 = 3$ m/day, $L_1 = 1000$ m, $L_2 = 500$m, $\rho_s = 1.025$ g/cm^3, $\rho_f =$ 1.000 g/cm^3, $w = 0.0005$ m/day, and $H_0 = 20$m.

Figure 15 shows how ground water travel times from a point at which x = 800m to the original coastline change with hydraulic conductivity of the reclamation materials for different reclamation lengths. After reclamation groundwater flow in the original aquifer tends to be slower so that the travel time becomes greater. This is because the reclamation increases the water table so that the water will flow through a cross-sectional area of greater height. The influence of the reclamation on groundwater travel times in the original aquifer diminishes gradually when K_2 increases and L_2 decreases, and becomes less sensitive to K_2 when K_2 is greater than 7 m/day. With L_2 = 500 m the change is 7.7 years for K_2 = 0.5 m/day and 0.9 years for K_2 = 10 m/day.

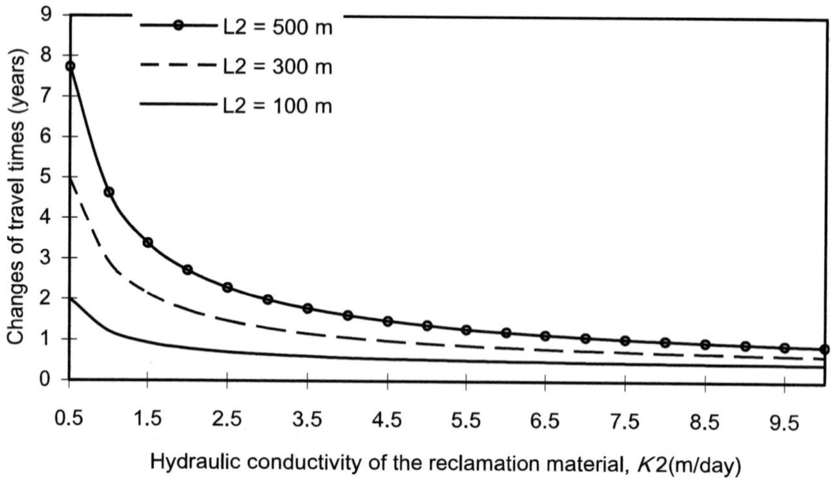

Figure 15. Changes of ground water travel times from a point at which x = 800m to the original coastline with different hydraulic conductivity of the reclamation materials for different reclamation lengths when K_1 = 3 m/day, L_1 = 1000 m, ρ_s = 1.025 g/cm^3, ρ_f = 1.000 g/cm^3, w = 0.0005 m/day, and H_0 = 20m.

Figure 16 shows the changes of ground water travel times from different positions in the original aquifer to the original coastline with distance from the water divide for different hydraulic conductivity of the reclamation materials when L_2=500m. Since the ground water in the whole original aquifer flows more slowly after reclamation, the increase of ground water travel times decreases with the distance from the ground water divide, which is similar to curves in Figure 14. When K_2 is low, the influence of the reclamation on groundwater flow will be more significant so that the changes of ground water travel times are greater. With K_2 = 1 m/day the change is 29.4 years for x = 200 m and 1.3 years for x = 950 m, indicating that changes of travel times are very sensitive to the distance towards the water divide.

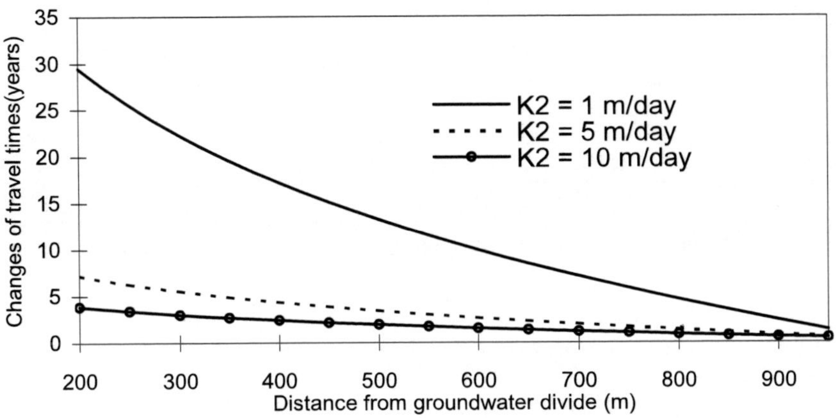

Figure 16. Changes of ground water travel times from different positions in the original aquifer to the original coastline with distance from the water divide for different hydraulic conductivity of the reclamation material when K_1 = 3 m/day, L_1 = 1000 m, L_2=500m, ρ_s = 1.025 g/cm^3, ρ_f = 1.000 g/cm^3, w = 0.0005 m/day, and H_0 = 20m.

2.3. Impact of Reclamation on Streamline Pattern and Flow Velocity

It is well known that recharge can cause the spreading of contaminants (e.g. plume diving), which should be considered when placing wells at sites. The streamlines originating at the locations for the source of contaminants can be used to decide how much spreading of contaminants to expect at a point down gradient from the source.

Streamline calculation

The calculation method will be applicable for a case of one-dimensional shallow flow in a water table aquifer receiving an infiltration of a constant rate along the entire upper boundary (Figure 17). The base of the aquifer is assumed to be horizontal and impermeable. Water entering the aquifer due to recharge along the water table would move down gradient along the streamline. In a Dupuit-type water table aquifer recharged by surface infiltration and discharged by a down-gradient, fixed-head boundary, the flux is related to the recharge, w, by

$$Q_x = w(x\text{-}x_0) \tag{43}$$

where Q_x is the discharge per unit width (L^2/T), and x_0 is the location of the upgradient water divide. The total discharge in the x direction per unit width, through a vertical cross section of height $b(x)$ is partitioned by the streamline originating at $x = x_s$ (Strack ,1984), which leads to

$$\frac{Q_x^{t}}{Q_x} = \frac{D(x)}{b(x)} = \frac{Q_x - Q_s}{Q_x} = \frac{w(x - x_s)}{w(x - x_0)} = \frac{x - x_s}{x - x_0} \tag{44}$$

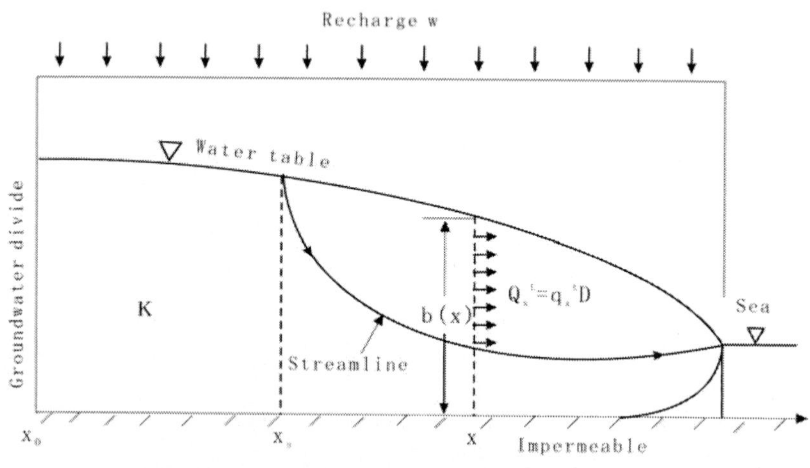

Figure 17. The streamline in an unconfined system recharged only by infiltration and discharged by a down-gradient, fixed-head boundary.

where Q_x^{t} is discharge above the streamline, Q_x is the total discharge, and $D(x)$ is the depth of the streamline below the water table. Note that (44) implies that

Dupuit assumption applies, i.e., equipotential surfaces are vertical and the flow is essentially horizontal. In other words, the water head *h(x)* and the specific discharge *q(x)* are constant over the height of the aquifer. Thus the depth of the streamline below the water table, *D(x)*, can be obtained as:

$$D(x) = \frac{x - x_s}{x - x_0} b(x) \tag{45}$$

Discussion of the analytical solutions

The streamlines can be calculated by combing Equations (30)-(31), (37) - (39), and (45). Assume that the saturated hydraulic conductivity K_1 of the aquifer is 0.1 m/day and the distance from the ground water divide to the original coastline, L_1, is 1000m. The densities of seawater and freshwater are 1.025 g/cm^3 and 1.000 g/cm^3, respectively. The infiltration rate is 0.0005 m/day and H_0 =20m. The hydraulic conductivity of the reclamation material, K_2, is 0.5 m/day and the reclamation length, L_2, is 500 m.

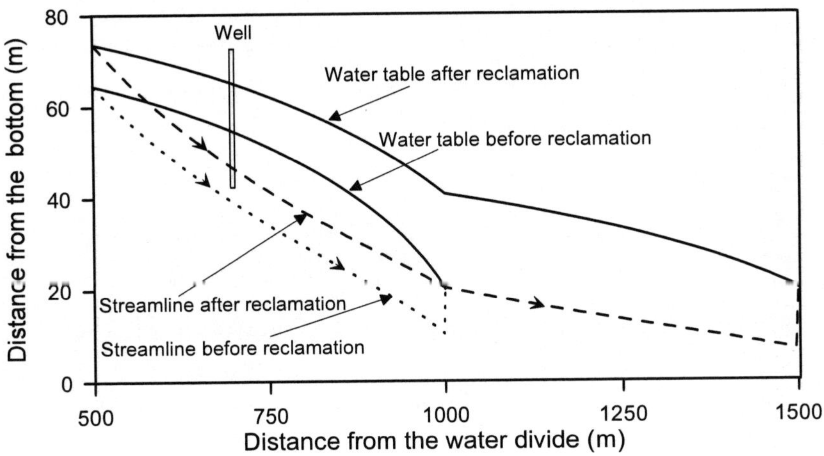

Figure 18. The water table and streamline originating at x = 500 m before and after reclamation, respectively (K_1=0.1m/day, K_2=0.5m/day, L_1=1000m, L_2=500m, ρ_s =1.025g/cm^3, ρ_f =1.000g/cm^3, w=0.0005m/day, and H_0=20m).

Figure 18 shows that how the water table and the streamline originating at x = 500m change before and after reclamation. One can see that the water table in the original aquifer increases significantly after reclamation. The streamline originating at x=500m also increases so that it may intersect the well (Figure 18). After reclamation the contaminants may enter some wells due to the increase of the elevation of the streamline originating from the source of contaminants, which should be considered when placing wells. One can also see that the recharged water will move across a vertical cross section of greater height so that the flow velocity becomes slower.

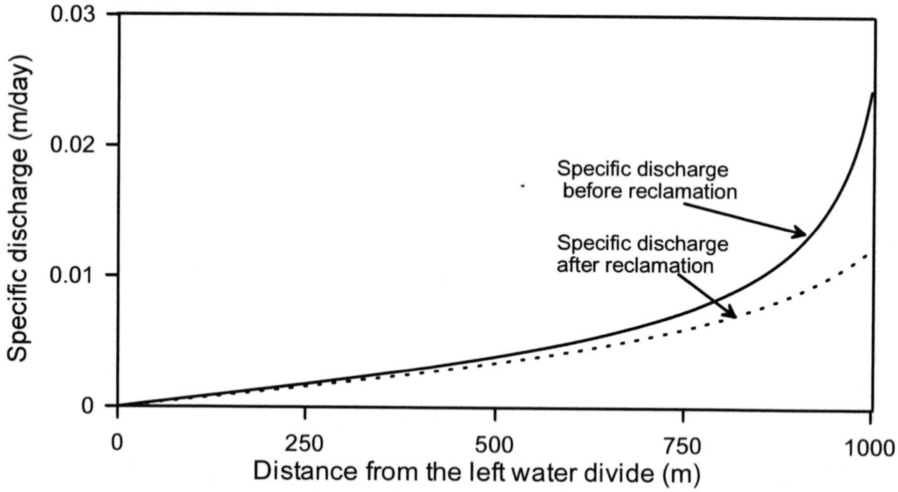

Figure 19. The specific discharge with distance from the water divide in the original aquifer before and after reclamation (K_1=0.1m/day, K_2=0.5m/day, L_1=1000m, L_2=500m, ρ_s =1.025g/cm^3, ρ_f =1.000g/cm^3, w=0.0005m/day, and H_0=20m).

Figure 19 shows how the specific discharge (Darcy flux) defined by Equation (25) changes before and after reclamation. The discharge per unit width will increase and the thickness of freshwater will decrease with distance from the water divide. Thus, as showed in Figure 19, the specific discharge increases with the distance from the water divide. After reclamation the height of the vertical cross section through which freshwater flows becomes greater because the reclamation increases the water table in the original aquifer. The discharge defined by Equation (43) remains unchanged so that the specific discharge decreases. As discussed previously, the maximal water level increase due to reclamation is at the original coastline, and the flow velocity at the water divide is always zero. Therefore, after reclamation the specific discharge becomes lower than that before reclamation especially in areas close to the original coastline.

Chapter 3

IMPACT OF LAND RECLAMATION IN AN ISLAND

When the reclamation scale is relatively small compared to the size of the original groundwater catchment, the previous assumption that the water divide remains unchanged (Figure 1) is valid. In the situation of an island the water divide may be moved when large-scale land reclamation occurs. In this case the reclamation on one side may change the ground water flow on both sides of the island.

3.1. IMPACT OF RECLAMATION ON GROUND WATER LEVEL AND SALTWATER INTERFACE

Two kinds of islands are considered, both with ground water flow resulting from precipitation recharge. In these two different kinds of islands, the unconfined flow system is bounded at the bottom by a horizontal impermeable layer (Figure 20) and by a saltwater-freshwater interface (Figure 27), respectively.

3.1.1. An Island Bounded Below by a Horizontal Impermeable Layer

3.1.1.1. Solution for Ground Water Level and Saltwater Interface

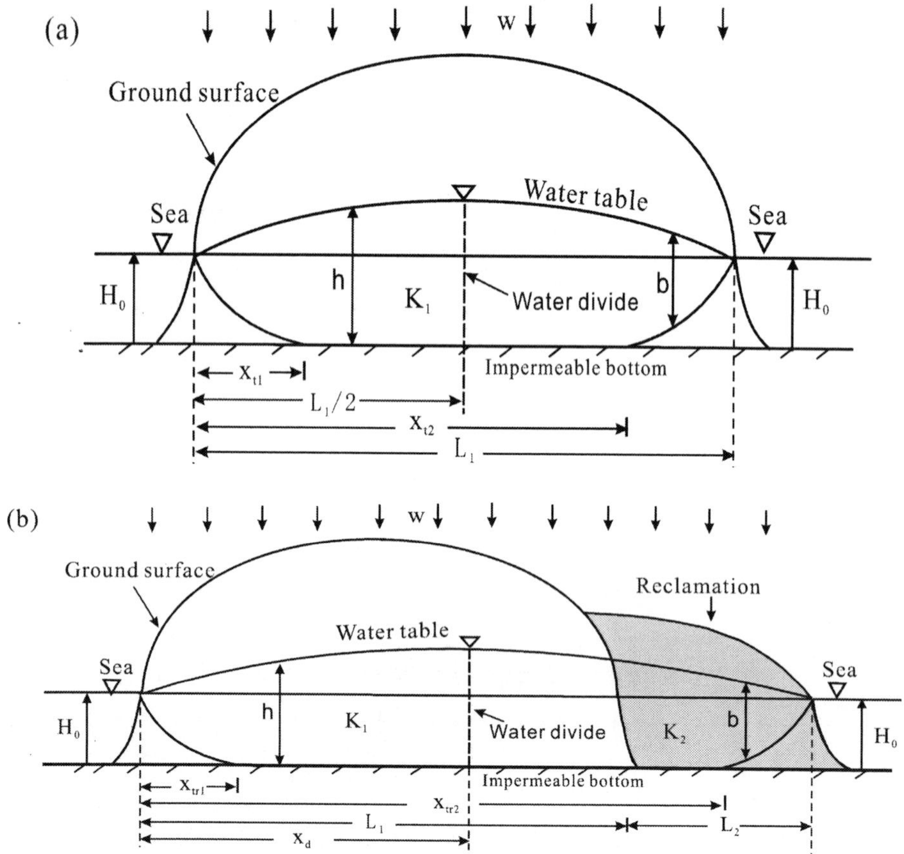

Figure 20. A schematic sketch of the ground water flow system in an island, which is bounded below by a horizontal impermeable layer and receives uniform vertical recharge (a) before reclamation and (b) after reclamation (Guo and Jiao, 2007).

Figure 20a shows an unconfined ground water system in a long island supported by a constant infiltration and bounded below by a horizontal impermeable layer. It can be a cross section through an elongated island, or a peninsula (Jiao et al. 2001). For convenience of discussion, it is referred to as a cross section in an island in the following discussion. The origin is located at the left coastline. L_1 is the width of the island before reclamation, and w is recharge rate per unit area. The distances from the left coastline to the tips of saltwater tongues are denoted as x_{t1} and x_{t2}, respectively. Guo and Jiao (2007) presented the solutions for the water levels which can be described as:

$$h = \sqrt{\frac{w(\rho_s - \rho_f)}{K_1 \rho_s}(L_1 x - x^2)} + H_0 \qquad (0 \le x \le x_{t1} \text{ or } x_{t2} \le x \le L_1) \tag{46}$$

$$h = \sqrt{\frac{w}{K_1}(L_1 x - x^2) + \frac{\rho_s}{\rho_f} H_0^2} \qquad (x_{t1} \le x \le x_{t2}) \tag{47}$$

At the tip of the saltwater tongue, the water head is equal to $\rho_s H_0 / \rho_f$ (see Equation 4). Set h equal to $\rho_s H_0 / \rho_f$ in either (46) or (47), one has

$$x^2 - L_1 x + \frac{K_1(\rho_s^2 - \rho_s \rho_f)}{w \rho_f^2} H_0^2 = 0 \tag{48}$$

The locations of the tips of saltwater tongues can be readily obtained by solving Equation (48) for x_{t1} and x_{t2}.

Figure 20b shows the unconfined ground water flow system in the island after reclamation. The hydraulic conductivity of the reclamation material and the reclamation length are denoted as K_2 and L_2, and the post-reclamation groundwater divide is located at $x = x_d$. After reclamation, the distances from the left coastline to the tips of saltwater tongues are represented as x_{tr1} and x_{tr2}, respectively. The solutions for the water level and the position of the ground water divide are (Guo and Jiao, 2007):

$$x_d = \frac{K_1 L^2 + (K_2 - K_1) L_1^2}{2(L_1 K_2 + L_2 K_1)} \tag{49}$$

$$h = \sqrt{\frac{w(\rho_s - \rho_f)}{K_1 \rho_s}(\beta x - x^2)} + H_0 \qquad 0 \le x \le x_{tr1} \tag{50}$$

$$h = \sqrt{\frac{w}{K_1}(\beta x - x^2) + \frac{\rho_s}{\rho_f} H_0^{\ 2}} \qquad x_{tr1} \le x \le L_1 \tag{51}$$

$$h = \sqrt{\frac{w}{K_2}(\beta x - x^2 + 2\gamma / w) + \frac{\rho_s}{\rho_f} H_0^{\ 2}} \qquad L_1 \le x \le x_{tr2} \tag{52}$$

$$h = \sqrt{\frac{w(\rho_s - \rho_f)}{K_2 \rho_s}(\beta x - x^2 + 2\gamma / w)} + H_0 \qquad x_{tr2} \le x \le L \tag{53}$$

where

$\beta = \left[K_1 L^2 + (K_2 - K_1) L_1^{\ 2} \right] / (L_1 K_2 + L_2 K_1)$ and

$\gamma = w L_1 L_2 L (K_2 - K_1) / (2 L_1 K_2 + 2 L_2 K_1)$.

Set h equal to $\rho_s H_0 / \rho_f$ in either (50) or (51), one has

$$x^2 - \beta x + \frac{K_1(\rho_s^{\ 2} - \rho_s \rho_f)}{w \rho_f^{\ 2}} H_0^{\ 2} = 0 \tag{54}$$

The distance from the left coastline to the tip of the saltwater tongue on the unreclaimed side, x_{tr1}, is one root of equation (54). Similarly, set h equal to $\rho_s H_0 / \rho_f$ in either (52) or (53), the location of the tip of the saltwater tongue on the reclaimed side, x_{tr2}, can be obtained as a root of the following equation:

$$x^2 - \beta x + \frac{K_2(\rho_s^2 - \rho_s \rho_f)}{w \rho_f^2} H_0^2 - 2\gamma / w = 0 \qquad (55)$$

Solving (54) and (55) for the tip locations is not difficult, but the process is cumbersome and is not shown here.

An important influence of the reclamation on the ground water flow system is that the water divide is displaced. The displacement of the water divide, Δd , can be calculated as

$$\Delta d = x_d - \frac{L_1}{2} = \frac{K_1 L_2 (L_1 + L_2)}{2(L_1 K_2 + L_2 K_1)} \qquad (56)$$

Equation (56) shows that the ground water divide will move toward the post-reclamation coastline, implying that ground water discharge to the sea on the left is increased. The displacement of the water divide is significant when K_1 is great. For a particular coastal area, the parameters L_1 and K_1 are fixed. In this case, Equation (56) implies that he displacement of the water divide, Δd , increases with L_2 and decrease with K_2. One can also see that the displacement of the water divide has nothing to do with the recharge rate w and the elevation of the sea level H_0.

3.1.1.2. Discussion of the Analytical Solutions Using a Hypothetical Example

A hypothetical example is employed to study the influence of the reclamation on ground water level and seawater-freshwater interface in an island bounded below by a horizontal impermeable layer. Assume that the width of the island before reclamation L_1 and the hydraulic conductivity of the original aquifer K_1 are 2000m and 0.1 m/day, respectively. The infiltration rate is taken as 0.0005 m/day and H_0 = 20m. The densities of seawater and freshwater are 1.025 g/cm^3 and 1.000 g/cm^3, respectively.

Change of the water level in the original aquifer

A comparison of predicted water levels at the original coastline on the reclaimed side (Figure 21) shows that the rise of the water level decreases with K_2 and increases with L_2. For a fixed value of L_2, the water level change tends to be sensitive to K_2 when K_2 is low. With L_2 = 500 m the change is 20 m for K_2 = 0.5 m/day and only 1.7 m for K_2 = 10 m/day, which indicates that the water level change can be very sensitive to K_2. The reclamation results in rise of the water table throughout the island (Figure 22), thereby increasing the volume of the freshwater. The water-table rise decreases with K_2 and increases with distance towards the left coastline. The shape of the water table in the original aquifer tends to be asymmetric after reclamation due to displacement of the water divide, as indicated by Equation (56). There is an obvious jump for the hydraulic gradient at the original coastline on the reclaimed side.

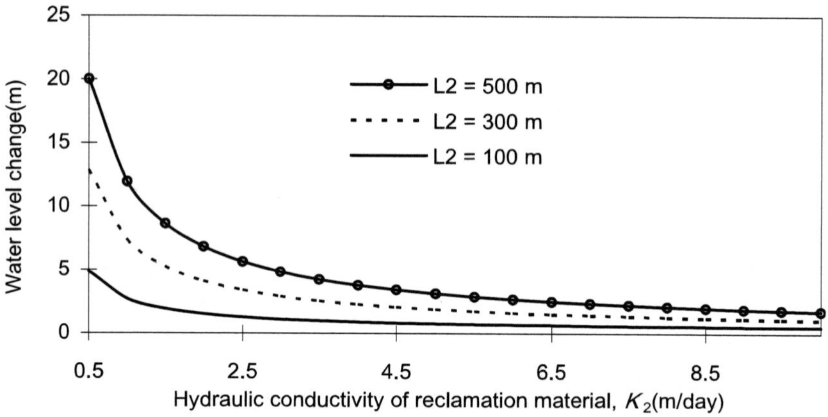

Figure 21. Change of the water level at the original coastline with hydraulic conductivity of the reclamation material for different reclamation lengths when K_1=0.1m/day, L_1=2000m, ρ_s =1.025g/cm^3, ρ_f =1.000g/cm^3, w=0.0005m/day, and H_0=20m.

Predicted displacement of the tip of the saltwater tongue on the reclaimed side as a function of the hydraulic conductivity of the reclamation material and reclamation length is presented in Figure 23, which shows the displacement of the tip decreases with K_2 and increases with L_2. After reclamation, the saltwater interface on the reclamation side is pushed toward the post-reclamation coastline, which in turns increases the volume of the freshwater. The displacement of the tip is similar to the results in Figure 7. For the sake of convenience, it is not presented here the influence of the reclamation on the saltwater tongue on the unreclaimed side. After reclamation, the water divide moves toward the reclaimed side, increasing the ground water discharge to the sea on the unreclaimed side. Thus, the saltwater interface on the unreclaimed side will also move seaward. Therefore, for a large-scale land reclamation project, the changes of ground water flow and seawater-freshwater interface over the whole island should be taken into account.

Usually, reclamation will increase the volume of the freshwater, a valuable water resource in the island situation.

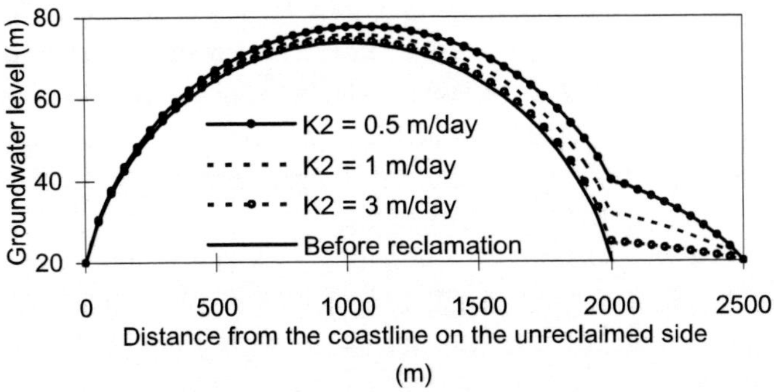

Figure 22. Water tables in the island with the distance from the coastline on the unreclaimed side for different hydraulic conductivity of the reclamation material when K_1=0.1m/day, L_1=2000m, L_2=500m, ρ_s =1.025g/cm^3, ρ_f =1.000g/cm^3, w=0.0005m/day, and H_0=20m (Guo and Jiao, 2007).

Displacement of the tip of the saltwater tongue

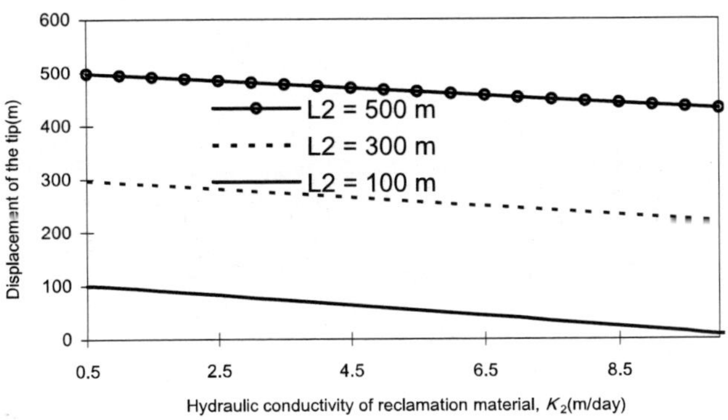

Figure 23. Displacement of the tip of the saltwater tongue on the reclaimed side with hydraulic conductivity of the reclamation material when the reclamation lengths are 500m, 300m and 100m, respectively (K_1=0.1m/day, L_1=2000m, ρ_s =1.025g/cm^3, ρ_f =1.000g/cm^3, w=0.0005m/day, and H_0=20m).

Change of the discharge to the sea on the unreclaimed side

Based on Equation (56), the percentage increase in flow to the sea on the unreclaimed side ($100 \times 2w\Delta d / wL_1 = 200\Delta d / L_1$) is plotted with the ratio of the reclamation length (L_2) to island width (L_1) (Figure 24). The percentage increase in flow increases with the ratio of the reclamation length to island width and decreases with K_2. With $L_2 = 500$ m, i.e., the ratio of the reclamation length to island width is 0.25, the percentage increase in flow to the sea on the unreclaimed side is 6% for $K_2 = 0.5$ m/day and 1% for $K_2 = 3$ m/day, indicating that the increase in discharge to the sea on the unclaimded side is very sensitive to K_2.

Figure 25 shows streamlines originating at x=500m and 1500m before and after reclamation. One can see that the elevations of streamlines on both sides of the island increase. On the unreclaimed side the streamline increases slightly more greatly than the water table so that the recharged water moves through a vertical cross section of lower height and the flow velocity of ground water becomes greater. On the reclaimed side, however, the water will flow more slowly because the increase of the streamline elevation is less than that of the water table.

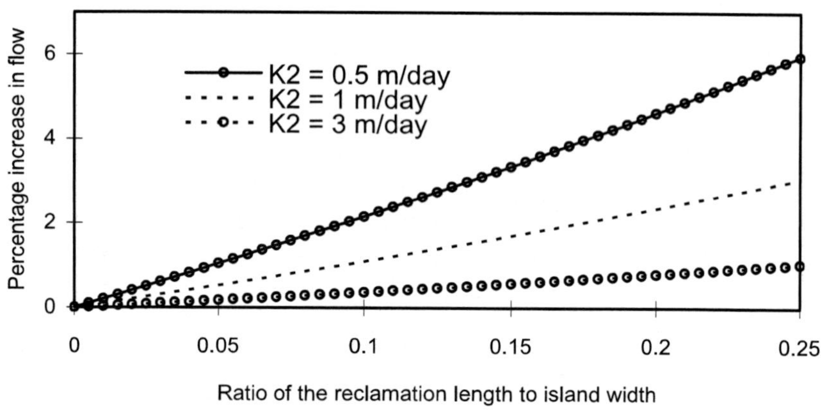

Figure 24. Percentage increase in the freshwater discharge to the unreclaimed side of the island versus the ratio of the reclamation length to island width for different K_2 when K_1=0.1m/day, L_1=2000m, ρ_s =1.025g/cm^3, ρ_f =1.000g/cm^3, w=0.0005m/day, and H_0=20m (Guo and Jiao, 2007).

Change of the streamline pattern and flow velocity

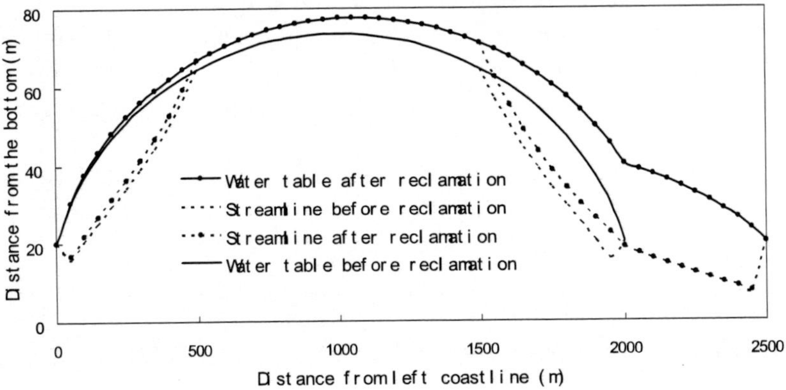

Figure 25. Change of the streamline pattern originating at x=500m and 1500m due to land reclamation with distance from the coastline on the unreclaimed side when $K_1 = 0.1$ m/day, $K_2 = 0.5$ m/day, $L_1 = 2000$ m, $L_2 = 500$ m, $\rho_s = 1.025$ g/cm^3, $\rho_f = 1.000$ g/cm^3, $w = 0.0005$ m/day, and $H_0 = 20$ m.

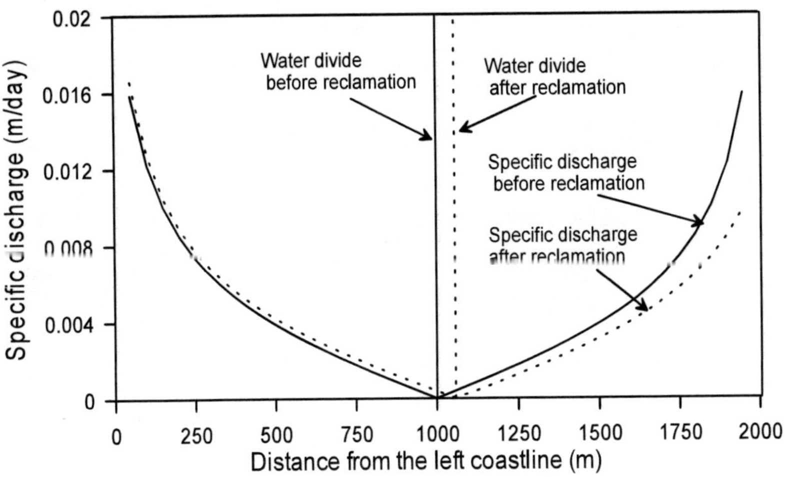

Figure 26. The specific discharge and the water divide in the original aquifer before and after reclamation (K_1=0.1m/day, K_2=0.5m/day, L_1=2000m, L_2=500m, ρ_s =1.025g/cm^3, ρ_f =1.000g/cm^3, w=0.0005m/day, and H_0=20m).

Figure 26 shows how the specific discharge and the water divide change in the original aquifer system before and after reclamation. Before reclamation the specific discharge decreases from the left coastline to the water divide and then increases gradually towards the coastline on the right. After reclamation, the specific discharge tends to increase on the unreclaimed side and decreases on the reclaimed side. The two curves intersect at a point in the area between the original water divide and the post-reclamation water divide. The influence of the reclamation on the flow velocity on the reclaimed side tends to be more significant than that on the unreclaimed side, and the influence increases with the distance from the left coastline.

3.1.2. An Island Bounded Below by Saltwater-freshwater Interface

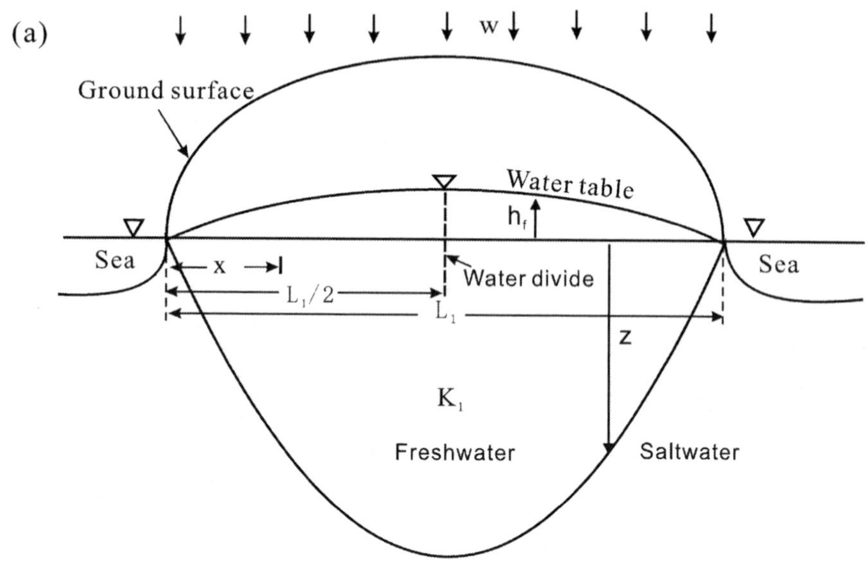

Figure 27. (Continued on next page.)

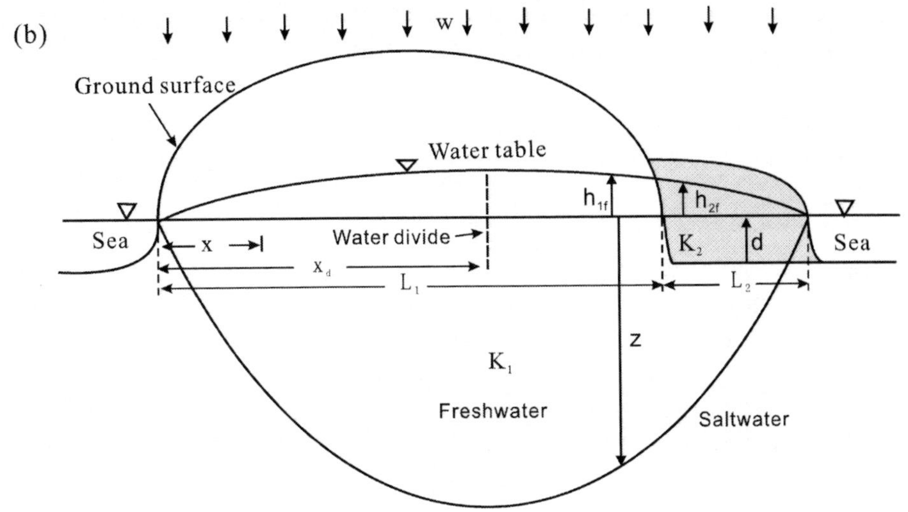

Figure 27. A schematic sketch of the ground water flow system beneath an oceanic island, which is bounded on the bottom by saltwater-freshwater interface and receives uniform vertical recharge : (a) before reclamation; (b) after reclamation.

3.1.2.1. Analytical Solution for Ground Water Level and Saltwater Interface

Figure 27a shows an approximate description of a long oceanic island supported by a constant rainfall infiltration rate. The freshwater lens, floating on top of sea water, is bounded above by a phreatic surface and below by a seawater-freshwater interface. The flow of the salt water beneath the interface may be neglected if the aquifer is thick. Here, x is the horizontal distance which origins at the left shoreline, L_1 is the original width of the island, w is the uniform vertical recharge rate per unit area, and K_1 is the hydraulic conductivity of the aquifer. The ocean surface is taken as the datum for the water head of fresh water $h_f(x)$, and $z(x)$ denotes the depth of the saltwater-freshwater interface below the sea level. The governing equation for freshwater flow based on Dupuit assumption and Ghyben-Herzberg relation is:

$$w(x - L_1/2) = -K_1(1+\delta)h_f \frac{dh_f}{dx} \tag{57}$$

where $\delta = \rho_f/(\rho_s - \rho_f)$. The boundary conditions are $h_f = 0$ at $x = 0$ (or L_1) and $dh_f/dx = 0$ at $x = L_1/2$. The solution to equation (57) is

$$h_f = \sqrt{\frac{w}{K_1(1+\delta)}(L_1 x - x^2)} \tag{58}$$

The expression (58) is identical to the solution presented by Henry (1964) and Fetter (1972).

Figure 27b shows the unconfined ground water flow system after reclamation. The reclamation length and the hydraulic conductivity of the reclamation material are denoted as L_2 and K_2, respectively. The water depth of the shallow sea is d. The water divide will shift to the right due to reclamation because ground water cannot discharge to the sea on the right as freely as it was before reclamation. Assuming that the water divide after reclamation is at $x = x_d$, the mathematical model can be described as:

$$w(x - x_d) = -K_1(1+\delta)h_{1f}\frac{dh_{1f}}{dx} \qquad (0 \le x \le L_1) \tag{59}$$

$$w(x - x_d) = -K_2(h_{2f} + d)\frac{dh_{2f}}{dx} - K_1(\delta h_{2f} - d)\frac{dh_{2f}}{dx}$$
$$(L_1 \le x \le L_1 + L_2) \tag{60}$$

with boundary conditions

$$h_{1f}(0) = 0 \tag{61}$$

$$h_{2f}(L_1 + L_2) = 0 \tag{62}$$

and the continuity conditions of ground water head and flux at the position where $x = L_1$ are

$$h_{1f}(L_1) = h_{2f}(L_1) \tag{63}$$

$q_1(L_1) = q_2(L_1)$, or

$$K_1(1+\delta)h_{1f}\frac{dh_{1f}}{dx} = K_2(h_{2f}+d)\frac{dh_{2f}}{dx} + K_1(\delta h_{2f}-d)\frac{dh_{2f}}{dx} \tag{64}$$

where h_{1f} and h_{2f} are the water head of fresh water in the original aquifer and the reclamation area, respectively. The solution of Equations (59) and (61) is

$$h_{1f} = \sqrt{\frac{w}{K_1(1+\delta)}(2x_d x - x^2)} \quad (0 \le x \le L_1) \tag{65}$$

The solution of Equations (60) and (62) is

$$\frac{K_2+\delta K_1}{2}h_{2f}{}^2 + d(K_2-K_1)h_{2f} = wx_d x - \frac{w}{2}x^2 + \frac{w}{2}(L_1+L_2)^2 - wx_d(L_1+L_2) \quad (L_1 < x \le L_1+L_2) \tag{66}$$

The following equation can be obtained by combining Equations (63), (64), (65) and (66):

$$\kappa t^2 + \nu t - (L_1L_2 + L_1{}^2) = 0 \tag{67}$$

where

$$t = \sqrt{\frac{w}{K_1(1+\delta)}(2x_d L_1 - L_1{}^2)}, \tag{68}$$

$$\kappa = \frac{L_1K_2 + L_2K_1 + \delta K_1(L_1+L_2)}{wL_2} \tag{69}$$

$$\nu = \frac{2dL_1(K_2-K_1)}{wL_2} \tag{70}$$

The solution to Equations (67) through (70) is

$$x_d = \frac{L_1}{2} + \frac{K_1(1+\delta)t^2}{2wL_1} \tag{71}$$

$$t = \frac{\sqrt{\nu^2 + 4\kappa(L_1L_2 + L_1^{\ 2})} - \nu}{2\kappa} \tag{72}$$

The solutions for ground water level and the seawater-freshwater interface after reclamation can be obtained with the combination of Equations (65), (66), (71) and (72). It should be noted that the ocean level is not involved in all these equations, indicating that the ocean level does not impact the shape and, thus, the volume of freshwater. However, as discussed previously, the volume of freshwater lens can be greatly impacted by the sea level if the island is bounded below by a horizontal impermeable bottom.

Equation (71) shows the water divide moves towards the coastline on the reclaimed side after reclamation, the change in the water divide is calculated as

$$\Delta d = x_d - \frac{L_1}{2} = \frac{K_1(1+\delta)t^2}{2wL_1} \tag{73}$$

The increased freshwater discharge to the sea on the unreclaimed side is calculated as:

$$w\Delta d = \frac{K_1(1+\delta)t^2}{2L_1} \tag{74}$$

3.1.2.2. Discussion of the Analytical Solutions Using a Hypothetical Example

A hypothetical example is employed to study the influence of the reclamation on the ground water flow pattern and the seawater-freshwater interface. Assume that the distance between the coastlines before reclamation L_1 and the hydraulic conductivity of the original aquifer K_1 are 2000m and 0.1 m/day, respectively. The infiltration rate is taken as 0.0005 m/day. The densities of seawater and freshwater are 1.025 g/cm^3 and 1.000 g/cm^3, respectively.

Change of the water level at the original coastline and displacement of the water divide

Figure 28 shows how the water level at the original coastline changes with K_2 for different reclamation lengths when the reclamation depth d is 20m. The water-level rise decreases with K_2 and increases with L_2. The water level buildup is very sensitive to the reclamation length and the hydraulic conductivity of the reclamation material. The increase of the water table is 9.3 m with $K_2 = 0.5$ m/day and $L_2 = 500$ m, and is only 0.3 m with $K_2 = 10$ m/day and $L_2 = 100$m. Comparing the results in Figure 28 with those in Figure 21, one can see that the influence of reclamation on water-level change in the original aquifer in an island bounded below by a horizontal impermeable layer is more significant than that in an island bounded below by seawater-freshwater interface.

Based on Equation (73), the displacement of the water divide (Δd) with hydraulic conductivity of the reclamation material is showed in Figure 29. The displacement becomes smaller as K_2 increases and L_2 decreases. The curves show that the displacement becomes more sensitive to L_2 and K_2 when K_2 becomes low, and the displacement of the water divide can be very significant. With $K_2 = 0.5$ m/day and $L_2 = 500$ m, the displacement of the water divide is 176.8 m, a great distance relative to the width of 2000 m of the island before reclamation.

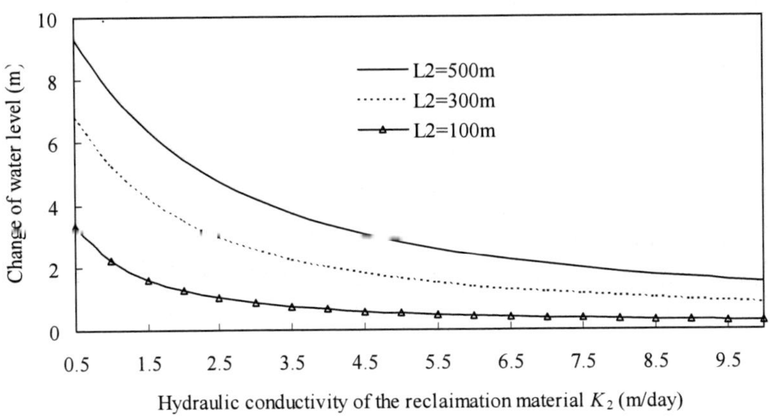

Figure 28. Change of the water level at the original coastline with the hydraulic conductivity of the reclamation material for different reclamation lengths when K_1=0.1m/day, L_1=2000m, ρ_s =1.025g/cm^3, ρ_f =1.000g/cm^3, w=0.0005 m/day, and d=20m.

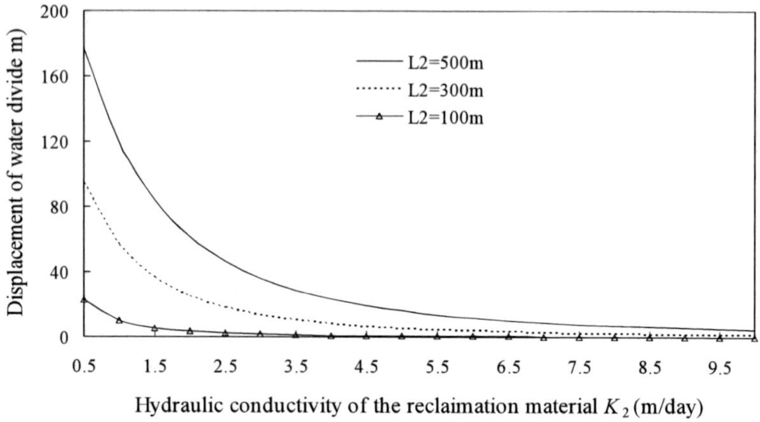

Figure 29. Displacement of ground water divide with hydraulic conductivity of the reclamation material for different reclamation lengths when K_1=0.1m/day, L_1=2000m, ρ_s =1.025g/cm^3, ρ_f =1.000g/cm^3, w=0.0005 m/day, and d=20m.

Change of the seawater-freshwater interface and the freshwater resource

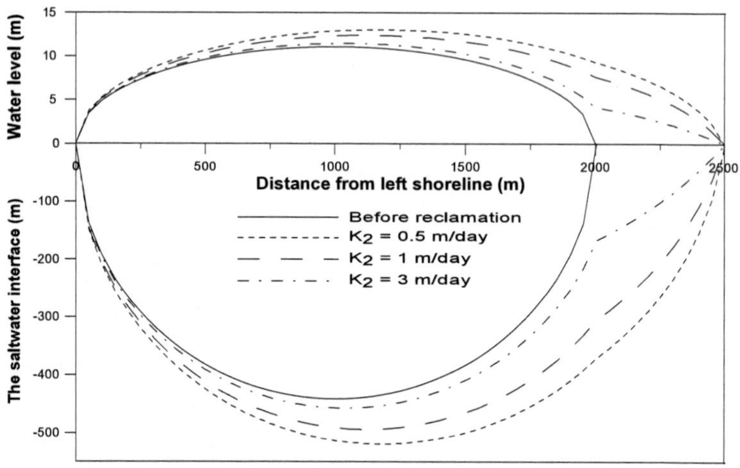

Figure 30. Changes of the water table and seawater-freshwater interface with distance from the coastline on the unreclaimed side when K_1 = 0.1 m/day, L_1 = 2000 m, L_2 = 500m ρ_s = 1.025g/cm^3, ρ_f = 1.000g/cm^3, w = 0.0005 m/day, and d = 20m.

Figure 30 shows changes of the water table (above the horizontal axis) and the seawater-freshwater interface (below the horizontal axis) with distance from the coastline on the unreclaimed side when L_2 and d are 500m and 20m, respectively. After reclamation, both the water table and the depth of the seawater-freshwater interface increase, thereby increasing the volume of the freshwater lens, an important resource in coastal areas. The increases in water table and the depth of the interface are more significant when K_2 decreases. The hydraulic gradient jumps at the interface between the original aquifer and the reclamation area due to the permeability contrast between the original aquifer and the reclamation material. When K_2 becomes greater, the permeability contrast will increase accordingly, thereby increasing the jump of the hydraulic gradient at the original coastline.

Figure 31 shows how the water table and the seawater-freshwater interface change with distance from the coastline on the unreclaimed side when L_2 and K_2 are 500m and 1m/day, respectively. The water table and the depth of the seawater-freshwater interface increase when the reclamation depth d decreases because K_2 (1m/day) is greater than K_1 (0.1m/day). If K_2 is lower than K_1, the water table and the depth of the seawater–freshwater interface can be expected to increase with the reclamation depth d.

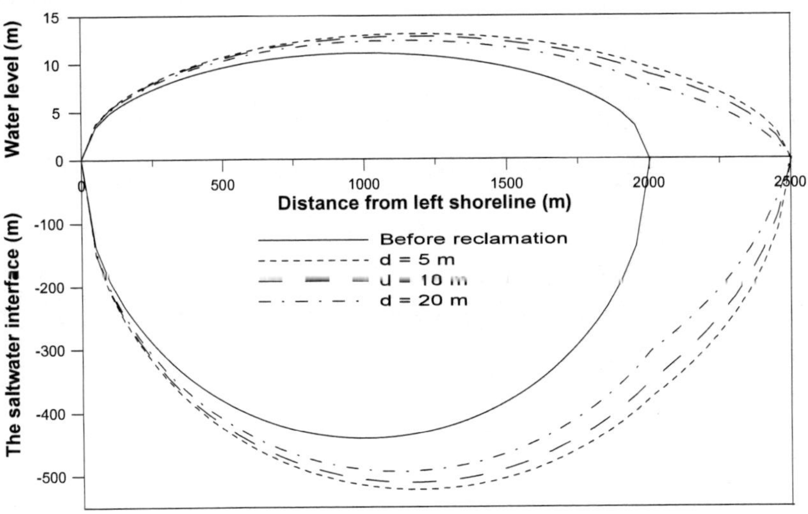

Figure 31. Change of the water table and the seawater-freshwater interface with distance from the coastline on the unreclaimed side when K_1 = 0.1 m/day, K_2 = 1 m/day, L_1 = 2000 m, L_2 = 500m ρ_s = 1.025g/cm^3, ρ_f = 1.000g/cm^3, w = 0.0005 m/day.

For simplicity, it is not presented here how the water table and the seawater-freshwater interface change with distance from the left coastline for different reclamation lengths. The previous analysis (Figure 28) shows that the water table and the depth of the seawater-freshwater interface will increase when the reclamation length L_2 becomes great. The volume of the freshwater lens (Figure 27b), therefore, is increased.

Change of the discharge to the sea on the unreclaimed side

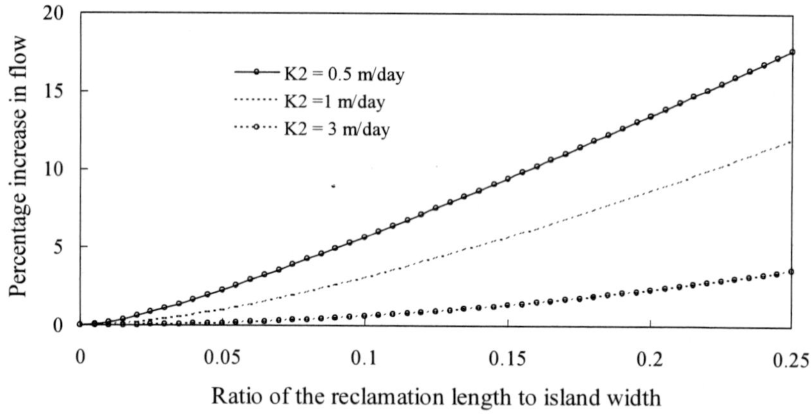

Figure 32. Percentage increase in the freshwater discharge to the unreclaimed side of the island versus the ratio of the reclamation length to island width for different K_2 when K_1=0.1m/day, L_1=2000m, ρ_s =1.025g/cm^3, ρ_f =1.000g/cm^3, w=0.0005 m/day, and d =20m.

Equation (74) shows that the freshwater recharge to the sea on the unreclaimed side will be increased. Using Equation (74), the percentage increase in the flow to the sea on the unreclaimed side ($100 \times 2w\Delta d / wL_1 = 200\Delta d / L_1$) is plotted with the ratio of the reclamation length (L_2) to island width (L_1) (Figure 32). The percentage increase in flow increases with the ratio of the reclamation length to island width and decreases with K_2, which is similar to the results in an island bounded below by an impermeable layer (Figure 24). When ratio of the reclamation length to island width is low and K_2 is great, the increase in flow to the left sea is minor and insensitive to K_2. The increase in flow, however, tends to be significant and sensitive to K_2 when ratio of the reclamation length to island width is great and K_2 is low, and increases almost linearly with ratio of L_2 to L_1.

Comparing the results in Figure 32 with those in Figure 24, one can see that the increase in discharge to the sea on the unreclaimed side, and thus the displacement of the water divide (see Equation 74), is more significant in an island bounded below by seawater-freshwater interface than that in an island bounded below by a horizontal impermeable layer. With L_2 = 500 m, i.e., the ratio of the reclamation length to island width is 0.25, the percentage increase in flow to the sea on the unreclaimed side is 6% in an island bounded below by a horizontal impermeable layer, and 17% in an island bounded below by seawater-freshwater interface.

Change of the streamline pattern and the flow velocity

Figure 33 shows the streamlines originating at x=500m and 1500m in the original aquifer before and after reclamation. After reclamation the elevation of the streamline on the unreclaimed side becomes higher but that on the reclaimed side becomes lower. The post-reclamation water table on the unreclaimed side will increase, but the rise of the water table is less than that of the streamline elevation (Figure 33). Therefore, the infiltrated water will move across a vertical cross section of lower height, which leads to a greater flow velocity. Similarly, the water will flow through a vertical cross section of greater height and the flow velocity becomes slower on the reclaimed side.

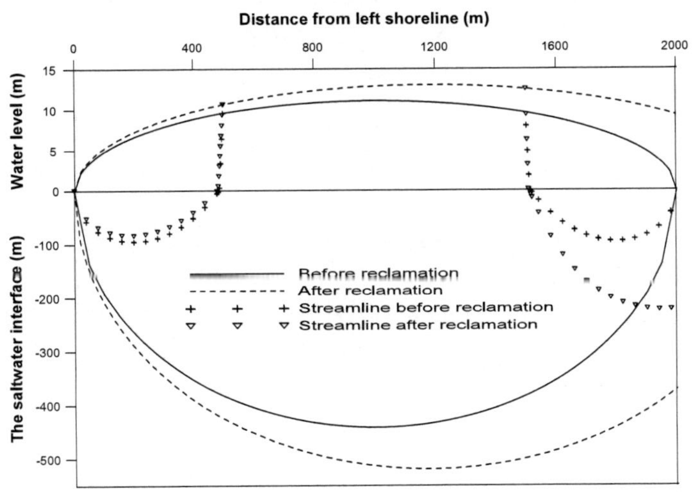

Figure 33. Change of streamlines originating at x=500m and 1500m in the original aquifer due to land reclamation with distance from the left shoreline when K_1 = 0.1 m/day, K_2 = 0.5 m/day, L_1 = 2000 m, L_2 = 500m ρ_s = 1.025g/cm^3, ρ_f = 1.000g/cm^3, w = 0.0005 m/day and d = 20 m.

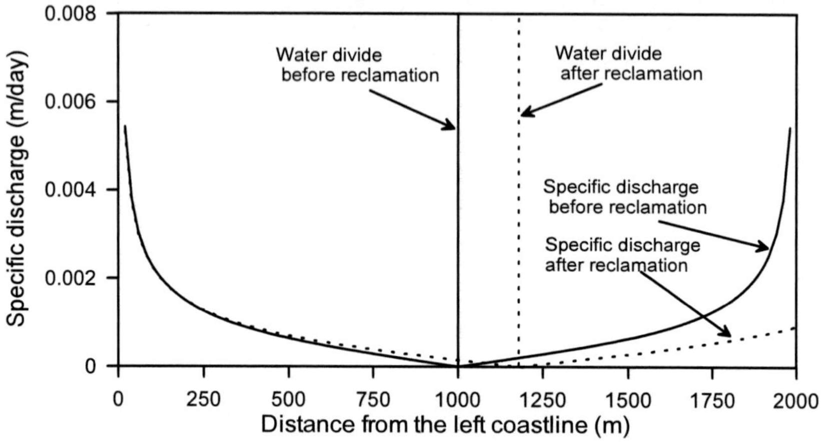

Figure 34. The specific discharge and the water divide in the original aquifer of the island before and after reclamation when $K_1 = 0.1$ m/day, $K_2 = 0.5$ m/day, $L_1 = 2000$ m, $L_2 = 500$m $\rho_s = 1.025\text{g/cm}^3$, $\rho_f = 1.000\text{g/cm}^3$, $w = 0.0005$ m/day and $d = 20$ m.

Figure 34 shows how the specific discharge and the water divide change in the original aquifer before and after reclamation. The specific discharge decreases from the coastline on the unreclaimed side to the water divide and then increases gradually towards the coastline on the right, which is similar to that in the island bounded below by an impermeable bottom (Figure 26). With the same parameter setting, the specific discharge (Figure 34) is much less than that in an island with a horizontal impermeable bottom. This is because the freshwater thickness is overall bigger so that the available surface for flow per unit width of aquifer is greater in an island bounded below by saltwater interface. After reclamation, the specific discharge tends to increase on the unreclaimed side and decrease on the reclaimed side, and the water divide moves towards the coastline on the reclaimed side. The displacement of the water divide is more significant than that in an island with a horizontal impermeable bottom (Figure 26) for the same parameter setting.

3.2. Impact of Land Reclamation on Ground Water Travel Time

In this section, analytical solutions are developed to estimate ground water travel times in unconfined aquifers located in an island bounded on the bottom by a horizontal impermeable layer, and the influence of the reclamation on ground water travel time is presented.

3.2.1. Analytical Solution

Before reclamation, Equation (34) remains valid in the portion between the water divide and the tip of the saltwater tongue. Equation (35) remains valid in the portion between the tip of the saltwater tongue and the corresponding coastline. In the coordinate system of the island setting, however, L_1 in Equations (34) and (35) should be replaced by $L_1/2$, x by $x - L_1/2$, and x_i by $x_i - L_1/2$. The solutions are:

$$\int_{t_i} dt = n_e \sqrt{\frac{\rho_s}{(\rho_s - \rho_f) w K_1}} \left[\sqrt{L_1 x - x^2} - \sqrt{L_1 x_i - x_i^2} - \frac{L_1}{2} \ln\left(\frac{L_1/2 + \sqrt{L_1 x - x^2}}{L_1/2 + \sqrt{L_1 x_i - x_i^2}} \frac{x_i - L_1/2}{x - L_1/2} \right) \right]$$

$$(0 \le x \le x_{t1} \;\; or \;\; x_{t2} \le x \le L_1) \qquad (75)$$

$$\int_{t_i} dt = n_e \sqrt{\frac{1}{wK_1}} \left[\sqrt{\chi^2 - (x - L_1/2)^2} - \sqrt{\chi^2 - (x_i - L_1/2)^2} - \chi \ln\left(\frac{\chi + \sqrt{\chi^2 - (x - L_1/2)^2}}{\chi + \sqrt{\chi^2 - (x - L_1/2)^2}} \frac{x_i - L_1/2}{x - L_1/2} \right) \right]$$

$$(x_{t1} \le x \le x_{t2}) \qquad (76)$$

where $\chi = \sqrt{L_1^2/4 + \frac{\rho_s K_1}{\rho_f w} H_0^2}$

Integrating Equations (75) and (76) with respect to x from x_0 to x_1 yields the travel time of water flow from one position ($x = x_0$) to another position ($x = x_1$) before reclamation. If $x_1 = 0$ (or L_1), the obtained travel time is the transit time of a particle of water originating at $x = x_0$.

After reclamation, Equation (34) remains valid in the portion between the water divide and the tip of the saltwater tongue on the unreclaimed side but with x_d in place of L_1 and x-x_d in place of x in the coordinate system of the island setting. Equation (35) remains valid in the portion between the tip of the saltwater tongue on the unreclaimed side and the corresponding coastline and the same substitutions of symbols are made. Similarly, Equation (40) remains valid in the portion between the water divide and the original coastline on the reclaimed side, Equation (41) remains valid between the original coastline and the tip of the saltwater tongue on the reclaimed side, and Equation (42) remains valid between the tip of the saltwater tongue and the coastline on the reclaimed side. However, the symbol L_1 should be replaced by $L_1 - x_d$ and x replaced by x-x_d. The final solutions are:

$$\int_{t_i}^{t} dt = n_e \sqrt{\frac{\rho_s}{(\rho_s - \rho_f) w K_1}} \left[\sqrt{2x_d x - x^2} - \sqrt{2x_d x_i - x_i^2} - x_d \ln\left(\frac{x_d + \sqrt{2x_d x - x^2}}{x_d + \sqrt{2x_d x_i - x_i^2}} \frac{x_i - x_d}{x - x_d} \right) \right]$$
$$(0 \le x \le x_{tr1}) \qquad (77)$$

$$\int_{t_i}^{t} dt = n_e \sqrt{\frac{1}{wK_1}} \left[\sqrt{\varepsilon^2 - (x - x_d)^2} - \sqrt{\varepsilon^2 - (x_i - x_d)^2} - \varepsilon \ln\left(\frac{\varepsilon + \sqrt{\varepsilon^2 - (x - x_d)^2}}{\varepsilon + \sqrt{\varepsilon^2 - (x_i - x_d)^2}} \frac{x_i - x_d}{x - x_d} \right) \right]$$
$$(x_{tr1} \le x \le L_1) \qquad (78)$$

$$\int_{t_i}^{t} dt = n_e \sqrt{\frac{1}{wK_2}} \left[\sqrt{\eta^2 - (x - x_d)^2} - \sqrt{\eta^2 - (x_i - x_d)^2} - \eta \ln\left(\frac{\eta + \sqrt{\eta^2 - (x - x_d)^2}}{\eta + \sqrt{\eta^2 - (x_i - x_d)^2}} \frac{x_i - x_d}{x - x_d} \right) \right]$$
$$(L_1 \le x \le x_{tr2}) \qquad (79)$$

$$\int_{t_i}^{t} dt = n_e \sqrt{\frac{\rho_s}{(\rho_s - \rho_f) w K_2}} \left[\sqrt{(L - x_d)^2 - (x - x_d)^2} - \sqrt{(L - x_d)^2 - (x_i - x_d)^2} \right.$$
$$\left. -(L - x_d) \ln\left(\frac{L - x_d + \sqrt{(L - x_d)^2 - (x - x_d)^2}}{L - x_d + \sqrt{(L - x_d)^2 - (x_i - x_d)^2}} \frac{x_i - x_d}{x - x_d} \right) \right]$$
$$(x_{tr2} \le x \le L) \qquad (80)$$

where $x_d = \left[K_1 L^2 + (K_2 - K_1) L_1^2 \right] / [2(L_1 K_2 + L_2 K_1)]$,

$$\varepsilon = \sqrt{x_d^{\ 2} + \frac{\rho_s K_1}{\rho_f w} H_0^{\ 2}}\ , \text{ and}$$

$$\eta = \sqrt{(L_1 + L_2 - x_d)^2 + \frac{\rho_s K_2 H_0^{\ 2}}{\rho_f w}}\ .$$

Integrating Equations (77), (78), (79) and (80) with respect to x from x_0 to x_1 yields the travel time of water flow from one position ($x = x_0$) to another position ($x = x_1$) in the unconfined aquifer after reclamation. Note that the water will flow towards the coastline on the reclaimed side when $x_0 > x_d$, and towards the coastline on the unreclaimed side when $x_0 < x_d$.

3.2.2. Discussion of the Analytical Solutions

Changes of ground water travel times due to the coastal reclamation are here discussed with a hypothetical example. Assume that the saturated hydraulic conductivity of the original aquifer is 3 m/day and the distance between the coastlines before reclamation, L_1, is 2000m. The infiltration rate is 0.0005 m/day and H_0 =20m. The effective porosity is assumed to be 0.3.

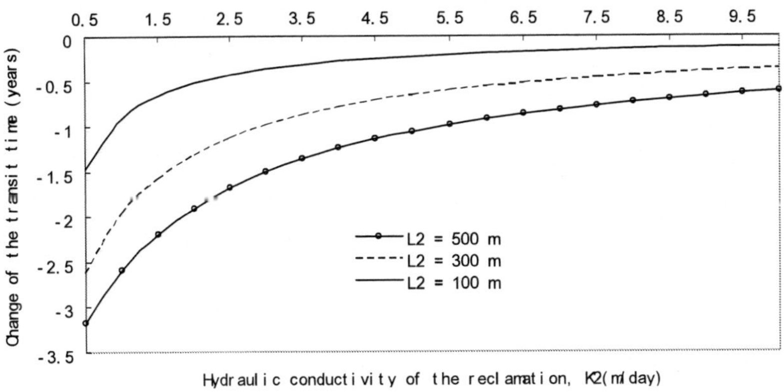

Figure 35. Changes of transit times from a point at x = 200m to the sea on the unreclaimed side with hydraulic conductivity of the reclamation material for different reclamation lengths when K_1 = 3 m/day, L_1 = 2000 m, ρ_s = 1.025g/cm^3, ρ_f = 1.000g/cm^3, w = 0.0005 m/day and H_0=20m.

The influence of reclamation on travel time on the unreclaimed side

Figure 35 shows changes of the transit times from a point at x = 200m to the sea on the unreclaimed side with hydraulic conductivity of the reclamation material for different reclamation lengths. One can see that values of changes of the transit times are negative, which indicates the water flows faster after reclamation. The change of transit times decreases with K_2 and increases with L_2 because the influence of the reclamation becomes more significant when K_2 decreases or L_2 increases. The decrease of the travel time is 3.2 years with K_2 = 0.5 m/day and L_2 = 500 m, and is only 0.12 years with K_2 = 10 m/day and L_2 = 100m. Change of the transit time tends to be low and insensitive to K_2 when K_2 is greater than 6m/day. Figure 36 shows how transit times to the sea from different positions in the original aquifer change with distance from the left coastline for different hydraulic conductivity of the reclamation material when L_2=500m. Changes of transit times increase with the distance from the left coastline, and tend to be more significant when K_2 is low. With x = 800m, the change is 31 years for K_2 = 1 m/day and only 9.9 years for K_2 = 10 m/day.

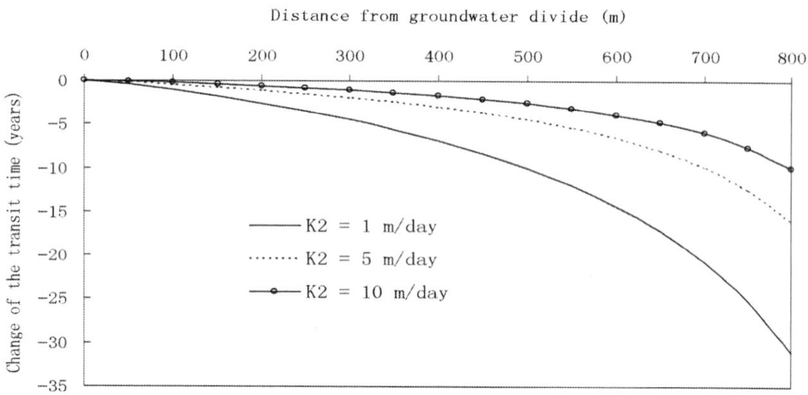

Figure 36. Changes of transit times to the sea on the unreclaimed side from different positions in the original aquifer with distance from the left coastline for different hydraulic conductivity of the reclamation materials when K_1 = 3 m/day, L_1 = 2000 m, L_2=500m ρ_s = 1.025g/cm^3, ρ_f = 1.000g/cm^3, w = 0.0005 m/day and H_0=20m.

The influence of reclamation on travel time on the reclaimed side

Figure 37 shows changes of travel times from a point at x = 1800 m to the original coastline on the reclaimed side with hydraulic conductivity of the reclamation material for different reclamation lengths. One can see that values of changes of travel times are positive, which indicates that the water flows more slowly after reclamation. The increase of the travel time decreases with K_2 and increases with L_2. Changes of travel times tend to be low and insensitive to K_2 when K_2 is greater than 5m/day. Comparision of Figures 35 and 37 shows that the influence of reclamation on ground water travel time is much more significant on the reclaimed side than that on the unreclaimed side. With K_2 = 0.5m/day and L_2 = 500 m, change of travel time over the last 200 m of the original aquifer is 72.8 years on the reclaimed side, and only 3.2 years on the unreclaimed side

Figure 38 shows how transit times from a point at x = 1800m to the sea on the reclaimed side change with hydraulic conductivity of the reclamation materials for different reclamation lengths. Ground water flow in the original aquifer tends to be slower so that the travel time becomes longer after reclamation (Figure 37). Also, the travel path of ground water becomes longer due to land reclamation. Therefore, transit times become greater than those before reclamation. The influence of the reclamation on groundwater transit time diminishes gradually when K_2 increases and L_2 decreases. For a fixed reclamation length, the impact of reclamation becomes less sensitive to K_2 when K_2 becomes great. Chang of transit times from a point at x = 1800 m can range from several years to more than one hundred years when K_2 and L_2 are combined differently. With K_2 = 0.5 m/day the change is 10.4 years for L_2 = 100 m and 123 years for L_2 = 500 m.

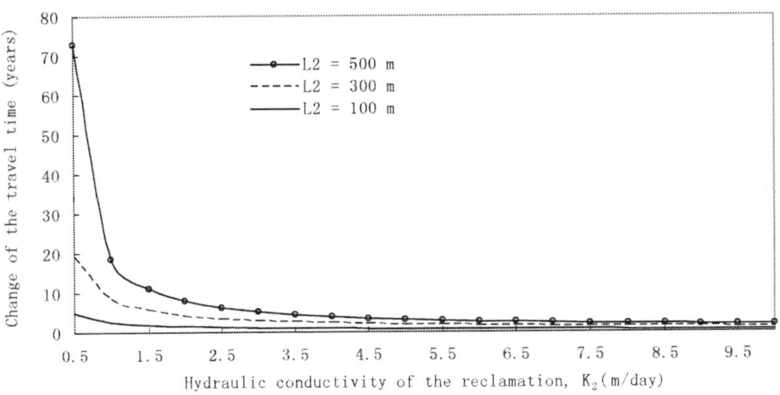

Figure 37. Changes of travel times from a point at x = 1800m to original coastline on the reclaimed side with hydraulic conductivity of the reclamation material for different reclamation lengths when K_1 = 3 m/day, L_1 = 2000 m, ρ_s = 1.025g/cm^3, ρ_f = 1.000g/cm^3, w = 0.0005 m/day and H_0 = 20 m.

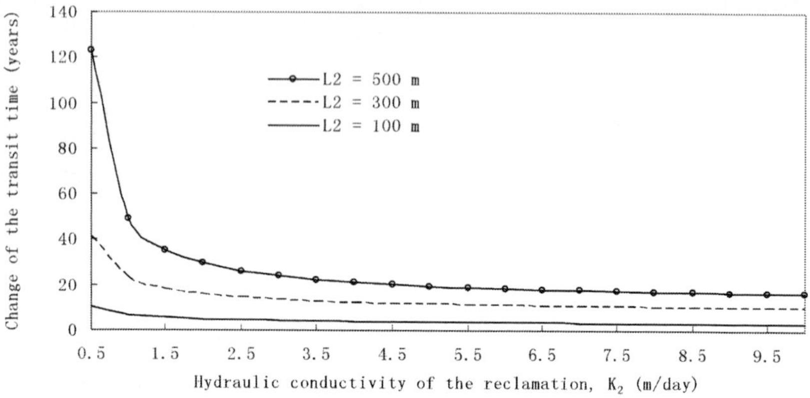

Figure 38. Changes of transit times from a point at x = 1800 m to the sea on the reclaimed side with hydraulic conductivity of the reclamation materials for different reclamation lengths when K1 = 3 m/day, L1 = 2000 m, ρ_s = 1.025g/cm3, ρ_f = 1.000g/cm3, w = 0.0005 m/day and H0=20m.

Chapter 4

SUMMARY AND CONCLUSIONS

The impact of land reclamation on ground water level and saltwater interface in an extensive land mass and an island are discussed. After reclamation, the water level in the original aquifer is increased and the saltwater interface moves seaward. The recharge can occur in a greater area so that the size of aquifer, and therefore the water resource, is increased. In the island situation, large-scale reclamation can result in the displacement of the water divide towards the reclaimed side, thereby increasing ground water discharge to the sea on both sides of the island. In all the cases, the impact of the reclamation on the water level and the saltwater interface mainly depends on the reclamation scale and the hydraulic conductivity of the reclamation material. In the case of an extensive land mass, the analytical solutions are checked against the results of a numerical code, which shows that only minor difference occurs.

Analytical solutions have been obtained to describe ground water travel time before and after reclamation for an extensive land mass and an island bounded below by a horizontal impermeable layer. Comparisons have been made between the solution in this paper and that developed by Chesnaux et al. (2005), which doesn't include the influence of seawater-freshwater interface. It is concluded that difference of these two solutions depends mainly on the elevation of the sea level, the infiltration rate and the hydraulic conductivity of the aquifer. The impact of reclamation on groundwater travel time is discussed detailedly based on the hypothetical examples. The groundwater travel time tends to be longer, and water flows more slowly in the original aquifer after reclamation. In the island situation, the ground water will flow more slowly in the original aquifer on the reclaimed side. However, the ground water travel time becomes shorter and water flows faster in the unreclaimed side. The impact of the coastal reclamation on ground water travel time is mainly controlled by the reclamation scale and hydraulic conductivity of the reclamation material. For the same reclamation scale and hydraulic conductivity of the reclamation material, the reclamation tends to have more significant impact on ground water flow on the reclaimed side than that on the unreclaimed side.

The impact of reclamation on flow velocity and streamline pattern is also presented. In some cases, the elevation of streamline increases after reclamation, which may impact spreading of contaminants and should be noted when placing wells. After reclamation the flow velocity will decrease especially in areas close to the original coastline for the situation of an extensive landmass. For the island setting, the flow velocity will decrease on the reclaimed side and increase on the unreclaimed side.

The analytical study in this paper is based on Dupuit and Ghyben-Herzberg assumptions. There exist some shortcomings in these assumptions such as no seepage face at the coastline and neglect of the vertical flow. In addition, the conceptual models considered here don not include the heterogeneity of the aquifer system, complicated topography at the coast, and the transient process immediately after the reclamation (Guo and Jiao, 2007). However, this study presents very useful insight into change of coastal ground water system in response to land reclamation.

APPENDIX A

A single-potential solution for regional interface problems in coastal aquifers was presented by Strack (1976), which was restricted to cases of steady flow with homogeneous isotropic permeability where the Dupuit assumption was valid. This approach is a more advanced and general method to handle combined unconfined flow, unconfined interface flow. In this Appendix, we describe the derivation of the analytical solution for groundwater level and seawater-freshwater interface in an extensive land mass in terms of the potential function instead of water head, and all symbols used are as defined previously.

SOLUTIONS FOR WATER LEVEL AND SALTWATER INTERFACE BEFORE RECLAMATION

Figure 1a shows a coastal unconfined ground water system receiving uniform vertical recharge. According to Strack (1976), the potentials for combined unconfined flow and unconfined interface flow are

$$\Phi = \frac{1}{2}K_1 h^2 + C_u \qquad h \geq \frac{\rho_s}{\rho_f}H_0 \tag{A1}$$

$$\Phi = \frac{1}{2}K_1 \frac{\rho_s}{\rho_s - \rho_f}(h - H_0)^2 + C_{ui} \qquad h \leq \frac{\rho_s}{\rho_f}H_0 \tag{A2}$$

$$C_{ui} - C_u = \frac{1}{2} K_1 \frac{\rho_s}{\rho_f} H_0^{\ 2} \tag{A3}$$

where ρ_f and ρ_s are the densities of fresh ground water and seawater. C_{ui} and C_u stand for two constants and the subscripts *ui* and *u* refer to unconfined interface flow and unconfined flow, respectively. Choosing $C_u = 0$, one obtains

$$C_u = 0 \qquad C_{ui} = \frac{1}{2} K_1 \frac{\rho_s}{\rho_f} H_0^{\ 2} \tag{A4}$$

The boundary conditions of the unconfined aquifer before reclamation (Figure 1a) are

$$x = 0 \qquad Q_x = -\frac{\partial \Phi}{\partial x} = 0 \tag{A5}$$

and

$$x = L_1 \qquad h = H_0 \qquad \Phi = \frac{1}{2} K_1 \frac{\rho_s}{\rho_f} H_0^{\ 2} \tag{A6}$$

where Q_x is the discharge vector defined as pointing in the direction of flow and having a magnitude equal to the discharge flowing through a surface perpendicular to the direction of flow, of unit width and of a height equal to the thickness of the flow region (Strack, 1976). The governing differential equation for the case of rainfall is

$$\nabla^2 \Phi = -w \tag{A7}$$

The solution of Equations (A5) through (A7) is

$$\Phi = \frac{1}{2} w(L_1^{\ 2} - x^2) + \frac{1}{2} K_1 \frac{\rho_s}{\rho_f} H_0^{\ 2} \tag{A8}$$

The heads in the aquifer can be obtained from (A1), (A2), and (A8):

$$h = \sqrt{\frac{w}{K_1}(L_1^2 - x^2) + \frac{\rho_s}{\rho_f} H_0^2} \qquad 0 \le x \le x_t \qquad \text{(A9)}$$

$$h = \sqrt{\frac{w(\rho_s - \rho_f)}{K_1 \rho_s}(L_1^2 - x^2) + H_0} \qquad x_t \le x \le L_1 \qquad \text{(A10)}$$

The expression for the location of the tip of the saltwater tongue can be obtained by setting h equal to $\rho_s H_0 / \rho_f$ in either (A1) or (A2) and then solving for x after substitution of this value of the potential in (A8), which yields

$$x_t = \sqrt{L_1^2 - \frac{K_1(\rho_s^2 - \rho_s \rho_f)}{w \rho_f^2} H_0^2} \qquad \text{(A11)}$$

SOLUTIONS FOR WATER LEVEL AND SALTWATER INTERFACE AFTER RECLAMATION

After reclamation (Figure 1b), the potentials in the reclaimed portion of the aquifer can be defined as

$$\Phi = \frac{1}{2} K_2 h^2 \qquad h \ge \frac{\rho_s}{\rho_f} H_0 \qquad \text{(A12)}$$

$$\Phi = \frac{1}{2} K_2 \frac{\rho_s}{\rho_s - \rho_f}(h - H_0)^2 + \frac{1}{2} K_2 \frac{\rho_s}{\rho_f} H_0^2 \qquad h \le \frac{\rho_s}{\rho_f} H_0 \qquad \text{(A13)}$$

The condition (A5) remains valid for the left boundary but the boundary condition at $x = L = L_1 + L_2$, is changed as:

$$x = L \qquad h = H_0 \qquad \Phi = \frac{1}{2} K_2 \frac{\rho_s}{\rho_f} H_0^2 \qquad \text{(A14)}$$

The governing equation for the potential throughout the flow region is the same as (A7). The solution is

$$\Phi=\frac{1}{2}w(L^2-x^2)+\frac{1}{2}K_2\frac{\rho_s}{\rho_f}H_0^{\ 2} \tag{A15}$$

If the hydraulic conductivity jumps at $x=L_1$ then (A15) remains valid, but the potential jumps:

$$\Phi^-/\Phi^+=K_1/K_2 \tag{A16}$$

In order to have a continuous potential, the potential in the original aquifer is redefined as

$$\Phi=\frac{1}{2}K_1h^2+\xi \tag{A17}$$

where $\xi=\Phi^+-\Phi^-=\dfrac{K_2-K_1}{K_2}\Phi^+$.

Now that the new potential is continuous throughout the flow region. The unknown value ξ can be obtained from (A15) by computing Φ at $x=L_1$:

$$\xi=\frac{K_2-K_1}{K_2}\Phi(L_1)=\frac{K_2-K_1}{2K_2}\left[w(L^2-L_1^{\ 2})+K_2\frac{\rho_s}{\rho_f}H_0^{\ 2}\right] \tag{A18}$$

The solution throughout the flow domain is now obtained. The head in the original aquifer can be computed from (A15) and (A17), and heads in the reclaimed portion of the aquifer can be obtained from (A12), (A13), and (A15):

$$h=\sqrt{w\left[\frac{1}{K_1}(L_1^{\ 2}-x^2)+\frac{1}{K_2}(2L_1+L_2)L_2\right]+\frac{\rho_s}{\rho_f}H_0^{\ 2}} \qquad 0\le x\le L_1 \tag{A19}$$

$$h = \sqrt{\frac{w}{K_2}\left[(L_1 + L_2)^2 - x^2\right] + \frac{\rho_s}{\rho_f} H_0^{\ 2}} \qquad L_1 \le x \le x_{tr} \tag{A20}$$

$$h = \sqrt{\frac{w(\rho_s - \rho_f)}{K_2 \rho_s}\left[(L_1 + L_2)^2 - x^2\right] + H_0} \qquad x_{tr} \le x \le L_1 + L_2 \tag{A21}$$

The expression for the location of the tip of the saltwater tongue can be obtained by setting h equal to $\rho_s H_0 / \rho_f$ in either (A12) or (A13) and then solve for x after substitution of the derived potential in (15), which yields

$$x_{tr} = \sqrt{(L_1 + L_2)^2 - \frac{K_2(\rho_s^{\ 2} - \rho_s \rho_f)}{w \rho_f^{\ 2}} H_0^{\ 2}} \quad . \tag{A22}$$

APPENDIX B

In this Appendix, we describe the derivation of the analytical solution for groundwater level and seawater-freshwater interface in an island, bounded below by a horizontal impermeable bottom, in terms of the potential function instead of water head, and all symbols used are as defined previously.

SOLUTIONS FOR WATER LEVEL AND SALTWATER INTERFACE BEFORE RECLAMATION

Figure 20a shows the unconfined ground water flow system before reclamation. The potentials in the whole flow region may be obtained by setting C_{ui} equal to 0 in (A1) and (A2), which yields

$$\Phi = \frac{1}{2}K_1 h^2 - \frac{1}{2}K_1 \frac{\rho_s}{\rho_f} H_0^{\ 2} \qquad h \geq \frac{\rho_s}{\rho_f} H_0 \qquad \text{(B1)}$$

$$\Phi = \frac{1}{2}K_1 \frac{\rho_s}{\rho_s - \rho_f}(h - H_0)^2 \qquad h \leq \frac{\rho_s}{\rho_f} H_0 \qquad \text{(B2)}$$

The boundary conditions of the unconfined system before reclamation are

$$h = H_0 \qquad \Phi = 0 \text{ when } x = 0 \text{ or } L_1 \qquad \text{(B3)}$$

The governing equation (A7) for the potential throughout the flow region remains valid. The solution is

$$\Phi=\frac{1}{2}w(L_1x-x^2) \tag{B4}$$

The heads in the aquifer may be obtained from (B1), (B2), and (B4):

$$h=\sqrt{\frac{w(\rho_s-\rho_f)}{K_1\rho_s}(L_1x-x^2)}+H_0 \qquad 0\le x\le x_{t1} \text{ or } x_{t2}\le x\le L_1 \tag{B5}$$

$$h=\sqrt{\frac{w}{K_1}(L_1x-x^2)+\frac{\rho_s}{\rho_f}H_0^{\,2}} \qquad x_{t1}\le x\le x_{t2} \tag{B6}$$

Set h equal to $\rho_s H_0/\rho_f$ in either (B1) or (B2) and then substitute this value of the potential in (B4), one has

$$x^2-L_1x+\frac{K_1(\rho_s^{\,2}-\rho_s\rho_f)}{w\rho_f^{\,2}}H_0^{\,2}=0 \tag{B7}$$

The locations of the tips of saltwater tongues can be readily obtained by solving for x_{t1} and x_{t2} in equation (B7).

SOLUTION FOR GROUND WATER LEVEL AND SALTWATER INTERFACE AFTER RECLAMATION.

Figure 20b shows the ground water flow system after reclamation. The potentials in the reclaimed portion of the aquifer are defined as

$$\Phi=\frac{1}{2}K_2h^2-\frac{1}{2}K_2\frac{\rho_s}{\rho_f}H_0^{\,2} \qquad h\ge\frac{\rho_s}{\rho_f}H_0 \tag{B8}$$

$$\Phi = \frac{1}{2} K_2 \frac{\rho_s}{\rho_s - \rho_f}(h - H_0)^2 \qquad h \le \frac{\rho_s}{\rho_f} H_0 \tag{B9}$$

If the hydraulic conductivity jumps at $x = L_1$ then the potential jumps:

$$\Phi^- / \Phi^+ = K_1 / K_2 \tag{B10}$$

In order to make the potential continuous throughout the flow region, the potentials in the original aquifer are redefined as

$$\Phi = \frac{1}{2} K_1 h^2 - \frac{1}{2} K_1 \frac{\rho_s}{\rho_f} H_0^{\,2} + \gamma \qquad h \ge \frac{\rho_s}{\rho_f} H_0 \tag{B11}$$

$$\Phi = \frac{1}{2} K_1 \frac{\rho_s}{\rho_s - \rho_f}(h - H_0)^2 + \gamma \qquad h \le \frac{\rho_s}{\rho_f} H_0 \tag{B12}$$

where $\gamma = \Phi^+ - \Phi^- = \dfrac{K_2 - K_1}{K_2} \Phi^+$.

The boundary conditions at $x=0$ and $x=L=L_1+L_2$, are

$$x = 0 \qquad h = H_0 \qquad \Phi = \gamma \tag{B13}$$

$$x = L \qquad h = H_0 \qquad \Phi = 0 \tag{B14}$$

The governing equation (A7) for the domain remains valid. The solution is

$$\Phi = -\frac{1}{2} w x^2 + \left(\frac{wL}{2} - \frac{\gamma}{L}\right) x + \gamma \tag{B15}$$

The unknown value γ can be obtained from (B15) by computing Φ at $x = L_1$:

$$\gamma = \frac{K_2 - K_1}{K_2}\Phi(L_1) = \frac{K_2 - K_1}{K_2}\left[-\frac{1}{2}wL_1^{\ 2} + (\frac{wL}{2} - \frac{\gamma}{L})L_1 + \gamma\right]$$

The solution is:

$$\gamma = \frac{wL_1L_2L(K_2 - K_1)}{2(L_1K_2 + L_2K_1)} \tag{B16}$$

The heads in the aquifer can be obtained from (B8), (B9), (B11), (B12), and (B15):

$$h = \sqrt{\frac{w(\rho_s - \rho_f)}{K_1\rho_s}(\beta x - x^2)} + H_0 \qquad 0 \le x \le x_{tr1} \tag{B17}$$

$$h = \sqrt{\frac{w}{K_1}(\beta x - x^2) + \frac{\rho_s}{\rho_f}H_0^{\ 2}} \qquad x_{tr1} \le x \le L_1 \tag{B18}$$

$$h = \sqrt{\frac{w}{K_2}(\beta x - x^2 + 2\gamma / w) + \frac{\rho_s}{\rho_f}H_0^{\ 2}} \qquad L_1 \le x \le x_{tr2} \tag{B19}$$

$$h = \sqrt{\frac{w(\rho_s - \rho_f)}{K_2\rho_s}(\beta x - x^2 + 2\gamma / w)} + H_0 \qquad x_{tr2} \le x \le L \tag{B20}$$

where $\beta = \left[K_1L^2 + (K_2 - K_1)L_1^{\ 2}\right]/(L_1K_2 + L_2K_1)$, and γ is obtained from (B16).

Set h equal to $\rho_s H_0/\rho_f$ in either (B11) or (B12) and then substitute the derived potential in (B15), one has

$$x^2 - \beta x + \frac{K_1(\rho_s^2 - \rho_s\rho_f)}{w\rho_f^2} H_0^2 = 0 \tag{B21}$$

The distance from the left coastline to the tip of the left saltwater tongue, x_{tr1}, is one root of equation (B21). Similarly, the location of the tip of the right saltwater tongue, x_{tr2}, can be obtained as a root of the follow equation:

$$x^2 - \beta x + \frac{K_2(\rho_s^2 - \rho_s\rho_f)}{w\rho_f^2} H_0^2 - 2\gamma / w = 0 \tag{B22}$$

ACKNOWLEDGEMENTS

The study was supported by the Research Grants Council of the Hong Kong Special Administrative Region, China (HKU 7105/02P).

REFERENCES

Amin, I. E. and Campana, M. E. (1996), A general lumped parameter model for the interpretation of tracer data and transit time calculation in hydrologic systems. *Journal of Hydrology*, **179**(1-4): 1-21.

Barnes RSK (1991), Dilemmas in the theory and practice of biological conservation as exemplified by British coastal lagoons, *Biological Conservation* **55**(3): 315-328.

Bear, J., A.H.-D. Cheng, S. Sorek, D. Ouazar, and I. Herrera, ed. (1999), *Seawater Intrusion in Coastal quifers-Concepts, Methods and Practices*. Dordrecht, The Netherlands: Kluwer Academic Publishers.

Bear, J., and G. Dagan (1964), Some exact solutions of interface problems by means of the hodograph method, *J. Geophys. Res.*, **69**(8): 1563-1572.

Code of Federal Regulations (1988), Title 10--Energy: Department of Energy (Part 960.2). Miamisburg, OH : LEXIS-NEXIS, Division of Reed Elsevier.

Chapuis, R. P., R. Chesnaux. (2006). Travel time to a well pumping an unconfined aquifer without recharge. *Ground Water* **44** (4): 600-603.

Chesnaux, RE, JW Molson, and RP Chapuis (2005), An analytical solution for ground water transit time through unconfined aquifers. *Ground Water*, **43**(4): 511 -517.

Cornaton F., Perrochet P. (2005), Groundwater age, life expectancy and transit time distributions in advective–dispersive systems; 1.Generalized reservoir theory. *Advances in Water Resources*, **29**(9), 1267-1291.

Diersch, H.-J.G.. (2002), FEFLOW finite element subsurface flow and transport simulation system reference manual, WASY Inst. for Water Resour. Plann. and Syst. Res., Berlin.

Ding Guoping (2006), *Impact of deep building foundations on coastal groundwater flow systems*, Ph.D thesis, Univ. of Hong Kong, China.

Etcheverry, D., and P. Perrochet (2000), Direct simulation of groundwater transit-time distribution using the reservoir theory, *Hydrogeol. J.*, **8**: 200-208.

Fetter, C. W. (1972), Position of the saline water interface beneath oceanic islands. *Water Resources Research* **8** (5): 1307-1315.

Fetter, C.W. (1994), *Applied Hydrogeology*. New York: Prentice-Hall.

Fórizs, I., and Deák, J. (1998), Origin of bank filtered groundwater resources covering the drinking water demand of Budapest, Hungary. In: *Application of isotope techniques to investigate groundwater pollution, IAEA-TECDOC-1046*, Vienna, pp.133-165.

Gelhar, L.W., and J.L. Wilson (1974), Ground-water quality modeling. *Ground Water* **12**(6): 399–408.

Glover, R.E. (1964), The pattern of fresh water flow in coastal aquifer. In Sea Water in Coastal Aquifers. *USGS Water-Supply Paper* 1613-C. Reston, Virginia: USGS.

Guo, H.P. and J.J. Jiao (2007), Impact of Coastal Land Reclamation on Ground Water Level and the Sea Water Interface. *Ground Water* 45 (3): 362–367.

Henry, H.R. (1964), Interfaces between salt water and fresh water in coastal aquifers. In Sea Water in Coastal quifers. USGS Water-Supply Paper 1613-C. Reston, Virginia: USGS.

Jiao, J. J. (2000), Modification of regional groundwater regimes by land reclamation. *Hong Kong Geologist* **6**, 29-36.

Jiao, J. J., S. Nandy, and H. Li. (2001), Analytical studies on the impact of reclamation on groundwater flow. *Ground Water* **39**(6): 912-920

Jiao, J.J. (2002), Preliminary conceptual study on impact of land reclamation on groundwater flow and contaminant migration in Penny's Bay, *Hong Kong Geologist,* **8**: 14-20.

Jiao, J. J., and Z. Tang. (1999), An analytical solution of groundwater response to tidal fluctuation in a leaky confined aquifer. *Water Resources Research* **35**(3): 747-751.

Jiao JJ, C Leung, KP Chen, JM Huang, and RQ Huang (2005), Physical and chemical processes in the subsurface system in the land reclaimed from the sea, in *Collections of Coastal Geo-Environment and Urban Development*, P399-407, China Dadi Publishing House, Beijing, China.

Jiao, J. J., XS Wang and S Nandy (2006), Preliminary assessment of the impacts of deep foundations and land reclamation on groundwater flow in a coastal area in Hong Kong, China, *Hydrogeology Journal*, **14** (1-2): 100-114.

Kolditz, O., R. Ratke, H.-J. G. Diershch, and W. Zielke (1998), Coupled Groundwater flow and. transport: 1. Verification of variable density flow and transport models. *Advances in water. resources* **21**: 27–46.

Lee J. Florea and Carol M. Wicks (2001), Solute transport through laboratory-scale karstic auifers. *Journal of Cave and Karst Studies*, **63**(2): 59-66.

Lumb P. (1976), Land reclamation in Hong Kong. In *Proceedings of Residential Workshop on Materials and Methods for Low Coast Road, Rail and Reclamation Works*, Leura, Australia, 299-314.

Mahamood, H. R., and D. R. Twigg (1995), Statistical analysis of water table variations in Bharain. *Quarterly Journal of Engineering Geology* **28**: s63-s74.

Nield, S.P., L. R. Townley, and A. D. Barr (1994), A framework for quantitative analysis of. surface water-groundwater interaction: Flow geometry in vertical section. *Water Resources. Research*, **30**(8): 2461-2475.

Ni JR, Borthwick AGL, and Qin HP (2002), Integrated approach to determining postreclamation coastlines, *Journal of Environmental Engineering-Asce*, **128** (6):
543-551.

Noske R. A. (1995), The ecology of mangrove forest birds in peninsular malaysia, *IBIS* 137 (2): 250-263.

Seasholes, N. S. (2003), *Gaining ground: a history of landmaking in Boston.* Cambridge, Mass. : MIT Press.

Simpson, M. J., T. P. Clement, and F. E. Yeomans (2003), Analytical model for computing residence times near a pumping well. *Ground water* **41**(3): 351-354.

Smith, A. J., and J. V. Turner (2001), Density-dependent surface water-groundwater interaction and nutrient discharge in the Swan-Canning Estuary, *Hydrol. Processes*, **15**, 2595-2616.

Smith, A. J. (2004), Mixed convection and density-dependent seawater circulation in coastal aquifers, *Water Resour. Res.*, **40**, W08309, doi:10.1029/2003WR002977.

Strack, O.D.L. (1976), A single-potential solution for regional interface problems in coastal aquifers. *Water Resources Research* **12**(6): 1165-1174.

Strack, O. D. L. (1984), Three-dimensional streamlines in Dupuit-Forchheimer models. *Water Resources Research* **20**(7): 812-822.

Stuyfzand, P. J. (1995), The impact of land reclamation on groundwater quality and future drinking water supply in the Netherlands. *Wat. Sci. Tech.* **31**(8): 47-57.

Suzuki, T. (2003), Economic and geographic backgrounds of land reclamation in Japanese ports. *Marine Pollution Bulletin* **47**:226-229.

Terawaki T, Yoshikawa K, Yoshida G, Uchimura M, and Iseki K (2003), Ecology and restoration techniques for Sargassum beds in the Seto Inland Sea, Japan, *Marine Pollution Bulletin* **47** (1-6): 198-201.

Townley, L.R., and M.R. Davidson (1988), Definition of a capture zone for shallow water table lakes, *Journal of Hydrology*, **104**: 53-76.

Varni, M., and J. Carrera (1998), Simulation of groundwater age distributions,. *Water Resour. Res.*, **34**: 3271–3281.

INDEX

A

B

C

R

S

T